THE INTERIOR WEST: A FIRE SURVEY

To the Last Smoke

SERIES BY STEPHEN J. PYNE

Volume 1, *Florida: A Fire Survey*
Volume 2, *California: A Fire Survey*
Volume 3, *The Northern Rockies: A Fire Survey*
Volume 4, *The Southwest: A Fire Survey*
Volume 5, *The Great Plains: A Fire Survey*
Volume 6, *The Interior West: A Fire Survey*

STEPHEN J. PYNE

THE INTERIOR WEST

A Fire Survey

THE UNIVERSITY OF
ARIZONA PRESS
TUCSON

The University of Arizona Press
www.uapress.arizona.edu

ISBN-13: 978-0-8165-3770-9 (paper)

Cover design by Leigh McDonald
Cover photo: *Soda Fire Sunset* by Hugo Sindelar

Library of Congress Cataloging-in-Publication Data
Names: Pyne, Stephen J., 1949– | Pyne, Stephen J., 1949– To the last smoke ; v.6.
Title: The interior West : a fire survey / Stephen J. Pyne.
Description: Tucson : The University of Arizona Press, 2017. | Series: To the last smoke ; volume 6 | Includes bibliographical references and index.
Identifiers: LCCN 2017042853 | ISBN 9780816537709 (pbk. : alk. paper)
Subjects: LCSH: Wildfires—Nevada—History. | Wildfires—Utah—History. | Wildfires—Colorado—History. | Wildfires—Nevada—Prevention and control—History. | Wildfires—Utah—Prevention and control—History. | Wildfires—Colorado—Prevention and control—History. | Forest fires—Nevada—History. | Forest fires—Utah—History. | Forest fires—Colorado—History. | Forest fires—Nevada—Prevention and control—History. | Forest fires—Utah—Prevention and control—History. | Forest fires—Colorado—Prevention and control—History.
Classification: LCC SD421.32.N3 P96 2017 | DDC 363.37—dc23 LC record available at https://lccn.loc.gov/2017042853

Printed in the United States of America
♾ This paper meets the requirements of ANSI/NISO Z39.48-1992 (Permanence of Paper).

For Sonja
old flame, eternal flame

❧

CONTENTS

Series Preface: To the Last Smoke *ix*
Preface to Volume 6 *xi*

Map of the Interior West 2

PROLOGUE: Arid Lands, Burning Lands 3

NEVADA: FROM ROTTEN BOROUGH TO BURNING MAN **13**
The Other Side of the Mountain: Washoe WUI 19
A Sink for Exotics 27
A Worthy Adversary 40
Mushroom Clouds 50
Deep Fire 57

OUTLIER: STRIP TRIP **69**

UTAH: ZION'S HEARTH **85**
CatFire: Wasatch WUI 93
Burning Bushes 100
Plateau Province 108

OUTLIER: MESA NEGRA **125**

COLORADO: ROCKY MOUNTAIN HIGHS, AND LOWS **133**
ColFire: The Front as Center 136
Firebugs 144
Then and Now, Now and to Come 150
Fatal Fires, Hidden Histories 156

EPILOGUE: The Interior West Between Two Fires 164

Note on Sources *169*
Notes *171*
Index *187*

Illustrations follow page *78*

SERIES PREFACE

To the Last Smoke

WHEN I DETERMINED to write the fire history of America in recent times, I conceived the project in two voices. One was the narrative voice of a play-by-play announcer. *Between Two Fires: A Fire History of Contemporary America* would relate what happened, when, where, and to and by whom. Because of its scope it pivoted around ideas and institutions, and its major characters were fires or fire seasons. It viewed the American fire scene from the perspective of a surveillance satellite.

The other voice was that of a color commentator. I called it *To the Last Smoke*, and it would poke around in the pixels and polygons of particular practices, places, and persons. My original belief was that it would assume the form of an anthology of essays and would match the narrative play-by-play in bulk. But that didn't happen. Instead the essays proliferated and began to self-organize by regions.

I began with the major hearths of American fire, where a fire culture gave a distinctive hue to fire practices. That pointed to Florida, California, and the Northern Rockies, and to that oft-overlooked hearth around the Flint Hills of the Great Plains. I added the Southwest because that was the region I knew best. The Interior West beckoned because I thought I knew its central theme and wanted to learn more about its margins. Then there were stray essays on places and themes that needed to be corralled into a volume, and there were all those relevant regions that needed at least token treatment. Some like the Lake States and Northeast no longer commanded the national scene as they once had, but their stories were interesting and needed recording, or like the Pacific Northwest or central oak woodlands spoke to the evolution of fire's American century in a new way. Alaska boasts its own regional subculture. I would include as many as possible into a grand suite of short books.

My original title now referred to that suite, not to a single volume, but I kept it because it seemed appropriate and because it resonated with my own relationship to fire. I began my career as a smokechaser on the North Rim of the Grand Canyon in 1967. That was the last year the National Park Service (NPS) hewed to the 10 a.m. policy and we rookies were enjoined to stay with every fire until "the last smoke" was out. By the time the series appears, 50 years will have passed since that inaugural summer. I no longer fight fire; I long ago traded in my pulaski for a pencil. But I have continued to engage it with mind and heart, and this unique survey of regional pyrogeography is my way of staying with it to the end.

Some funding for the project came from the U.S. Forest Service, Department of the Interior, and Joint Fire Science Program, part of the residual monies left after researching *Between Two Fires* and the early volumes of *To the Last Smoke*. I'm grateful for their support. And of course the University of Arizona Press deserves praise as well as thanks for seeing the resulting texts into print. A special acknowledgement, too, goes to Kerry Smith, who continues to copyedit the series with precision and good sense.

PREFACE TO VOLUME 6

OVER THE SUMMER of 2016 I began a series of interviews and road trips to Nevada, Utah, and Colorado. They all, to my mind, belonged to what I considered the Interior West. They united around a shared aridity. They divided around profoundly different landscapes, biotas, and human histories. But they also promised a geodetic marker by which to triangulate American fire history because the first scientific inquiry into landscape fire in the United States had its origins here.

I thought I knew something about the region, at least in parts, and was eager to understand more of the rest. The reconnaissance would return me to themes, places, and personalities I had researched for my first book (and doctoral dissertation), a biography of G. K. Gilbert. In fact, my travels turned out to be far more difficult to schedule than I had anticipated or had experienced in my other surveys. The old research template broke down. What, elsewhere, had been stops along a continuous trek here became a halting series of broken journeys full of improvisations. The region disintegrated into state-based subregions, each with its own minisurvey, though they shared themes like front-range urbanization and problem species. The upshot is that these essays are paradoxically both more scattered and more deliberately designed—less the outcome of free associations with the sites I visited—than the others in the series. Moreover, since the essays need to stand alone as well as join an ensemble, new ones often reflected back on old ones or repeated common topics. I was unable to dampen all the echoing that resulted.

Those who made my visit productive (and, in some cases, possible) are acknowledged in the individual essays. But I offer my collective thanks again here.

THE INTERIOR WEST:
A FIRE SURVEY

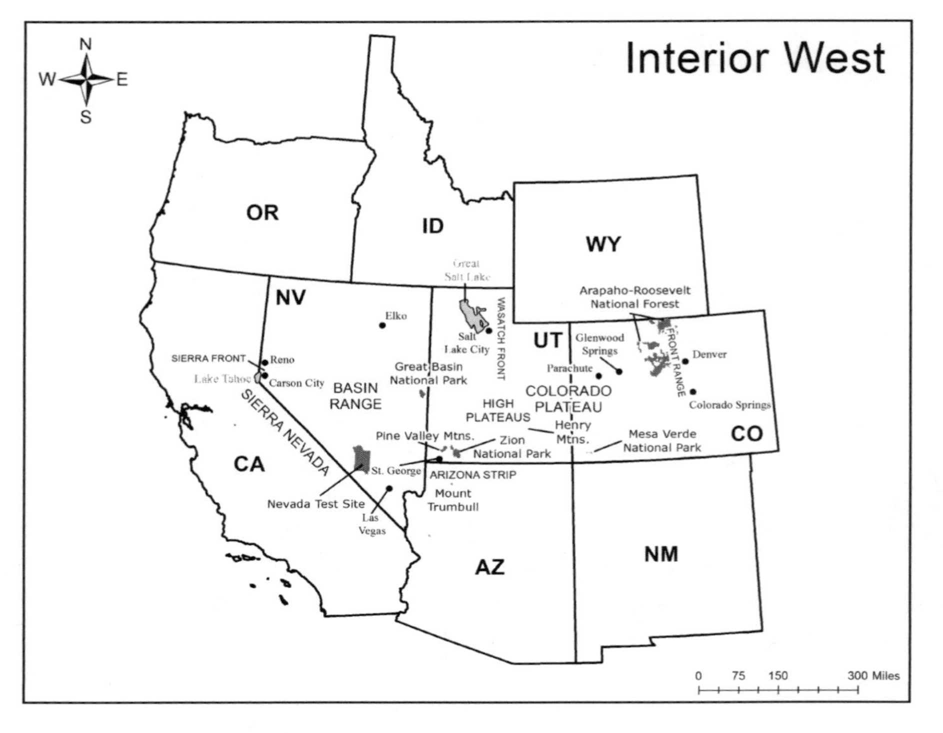

Map of the Interior West region

PROLOGUE

Arid Lands, Burning Lands

FOR MOST OF modern American fire history the Interior West has been the center that didn't hold. It was the hole in the national fire donut. The sweep of prairie fires broke against the Rockies. The famed big blowups that electrified agencies like the Forest Service ranged to the north, following the Rockies as they bowed north and west. The great firefights that defined postwar America broke out west of the Sierras. What the center had were brief outbursts in the mountains and occasionally larger and longer burns in the sagebrush steppe, though no one much cared.

There were plenty of fires—they came with the spring drought and ended with the winter snow. But they passed like a summer thunderstorm, occasionally violent but with no lasting consequence. They shaped no national policies, as the big burns in the Northern Rockies and Southern California did. They held no fire traditions that might inspire a national debate, as Florida had with prescribed fire, or even regional lore as the Flint Hills burners had. The intermountain interior boasted no charismatic megafires, no mesmerizing Big Blowups, no monumental fire sieges, and until 1994 no fire that commanded national attention. (Even the 1964 Elko fires affected the fledgling Bureau of Land Management [BLM], not national fire politics.) No school of fire science specialized in the ecology and history of its fire scene. In the pyrogeography of contemporary America, those flames were invisible. You didn't advance a career by spending years amid its remote forests, stunted mesas, or bleached sagebrush. The national action bordered the region's fringe provinces. The center itself was the Great Empty.

An explanation begins with geology and climate.

The middle of the American West resembles a great tectonic bubble bounded on the east by the Rocky Mountains and on the west by the Sierra Nevada, with the Wasatch fault dividing the scene roughly in half. It's as though the American cordillera had developed a geologic aneurism, swelled, and then burst.

What remained are two terrains that invert each other, one a hump, the other a hole. The hump is the Colorado Plateau, a colossal dome rippled with mesas, gouged with gorges, and spiked with occasional volcanic peaks. The hole is a hydrographic sink, the Great Basin, the land broken into blocks of parallel trending mountains and valleys—the Basin Range—as though an immense arch had fallen into rubble. The Colorado Plateau boasts the largest concentration of national parks and monuments in the United States, the Great Basin the least among public lands.

What they share is aridity, a seasonal rhythm of wetting and drying, wetter in some places, drier in others, given to odd years of drought and deluge. It's a good formula for fire, a poor one for American-style settlement. The Interior West is formidable country, not easily inhabited, better suited for mining camps and ski resorts than rooted villages and arable farms. Only along the watered Wasatch Front did westering Americans successfully colonize with something like permanency.

Yet geography is not destiny. Paradoxically, the perception of emptiness both mismaps the region's contemporary fire scene and misreads America's fire history.

In recent years the region is aflame with fires that grab national attention. Its three framing mountain ranges have become critical sites for the wildland-urban interface (WUI), literal flaming fronts in the spread or more accurately the sprawl of a combustion contagion. What had been a California quirk has leaped to Colorado Springs, Denver, and

Boulder; Reno, Sparks, and Carson City; and the chain of settlements down the Wasatch, twisting like tendrils from the Salt Lake Valley. The playas have become battlegrounds for endangered habitats as juniper and cheatgrass overrun sagebrush. The isolated plateaus and peaks have been ravaged by beetles and after fires by floods.

With those fires has come a reevaluation of significance. While the apexes of America's fire triangle continue to do more vigorously what they have always done, the interior is changing. Colorado is to the millennial era what California was to the postwar era, and rivals California for houses burned. Nevada has morphed from fire's salt playa into a poster child of a fire regime where habitat, wildfire, and global change are colliding, spiraling into a bottomless sink for Department of the Interior fire funds. Utah threatens to exchange its traditional beehive for a burned snag, overflowing with fires as the Great Salt Lake overflowed its shoreline in the early 1980s.

So, too, the perception of a regional fire void also misconstrues American fire history. The interior's center, Utah, was the scene for the country's first scientific inquiry into fire, the source of its first map of burned areas, and the inspiration for its first debate over fire policy and the role of government in executing it. All of that flowed from the Geographical and Geological Survey of the Rocky Mountain Region under the direction of Major John Wesley Powell. In 1878 it published, as the summary of its research, the *Report on the Lands of the Arid Region of the United States*, arguably the originating document for a science-supported, state-sponsored strategy of conservation.

The *Arid Lands* report included a map that classified Utah lands by four categories: desert, forest, irrigable, and burned. That established a model for C. S. Sargent's national map of forest fires for the 1880 census. The year the *Arid Lands* report was reprinted, 1879, Congress merged the various western surveys into a U.S. Geological Survey (USGS); Powell served as its second director, from 1881 to 1894. His successor, Charles Walcott, used the Survey's early experience in Utah as a template for the massive inventory of forest reserves (and their fires) conducted by the USGS after the 1897 organic act for managing the new forest reserves. Paradoxically, Utah, so seemingly out of step with American mores, provided the inspiration that evolved into Progressive Era conservation reforms.

Much as old burns help determine future ones, so a century after the Powell Survey, the region provided one of the prototypes for Landfire, a national inventory of land cover and fire. If *Arid Lands* identified land use suitable for 19th-century frontier settlement, Landfire helped reclassify America's national estate according to criteria relevant to the late 20th century.[1]

With the map came an extended caption and, after that commentary had circulated for a decade, a debate, which continues to inform the American fire scene. It's worth pausing to reconsider that moment because it reminds us that our national fire history was not inevitable, that there were alternatives present at the creation.

The nation's inaugural map of fires followed its first reconnaissance of a regional fire scene by credentialed scientists and its first assessment of fire and policy in the West; these became models for the great transition from laissez-faire pioneering to state oversight of natural resources. For years Powell and his corps—two of whom, G. K. Gilbert and Clarence E. Dutton, were subsequently elected to the National Academy of Sciences—had surveyed Utah. They appreciated that permanent settlement required irrigation, that irrigation depended on secure watersheds, that those watersheds lay in the forested mountains and high plateaus, and that those forests were threatened by fire and axe, of which fire seemed the greater menace. Where there were dry seasons, fires were possible, and where people were busy remaking habitats, fires were inevitable.

How did they see fire on the land? "The timber regions are only in part *areas of standing timber* [italics in the original]. This limitation is caused by fire. Throughout the timber regions of all the arid land fires annually destroy larger or smaller districts of timber, now here, now there, and this destruction is on a scale so vast that the amount taken from the lands for industrial purposes sinks by comparison into insignificance." There was ample timber, and adequate renewal by annual growth. There was no need to restrict access to it for local use, if—the critical "if"—fire could be controlled.[2]

The topic sentence followed: "*The protection of the forests of the entire Arid Region of the United States is reduced to one single problem—Can these forests be saved from fire?*" (italics added).[3]

> The writer has witnessed two fires in Colorado, each of which destroyed more timber than all that used by the citizens of that State from its settlement to the present day, and at least three in Utah, each of which has destroyed more timber than that taken by the people of the territory since its occupation. Similar fires have been witnessed by other members of the surveying corps. Everywhere throughout the Rocky Mountain Region the explorer away from the beaten paths of civilization meets with great areas of dead forests: pines with naked arms and charred trunks attesting to the former presence of this great destroyer. The younger forests are everywhere beset with fallen timber, attesting to the vigor of the flames, and in seasons of great drought the mountaineer sees the heavens filled with clouds of smoke.[4]

But if its arid climate leveraged sparks into free-ranging fires, the fact remained that people cast most of those sparks. Powell placed responsibility squarely onto the native peoples. "In the main these fires are set by Indians. Driven from the lowlands by advancing civilization, they resort to the higher regions until they are forced back by the deep snows of winter." Want caused by the loss of traditional lands, desire prompted by "luxuries to which they were strangers in their primitive condition," ability augmented by new technologies—these led to continued and expanded burning. "On their hunting excursions they systematically set fire to forests for the purpose of driving game. This is a fact well known to all mountaineers. Only the white hunters of the region properly understand why these fires are set, it being usually attributed to a wanton desire on the part of the Indians to destroy that which is of value to the white man. The fires can, then, be very greatly curtailed by the removal of the Indians."[5]

The scene that the Powell Survey mapped was, then, one upset by colonization. The old regime, which also included native burning but in somewhat different ways, was passing. "To preserve the forests they must be protected from fire. This will be largely accomplished by removing the Indians." Eventually protection will come, though it will be "tardy, for it depends upon the gradual settlement of the country; and this again depends upon the development of the agricultural and mineral resources and the establishment of manufactories, and to a very important extent on the building of the railroads." Lightning fire did not appear significant since the obviously damaging burns were set by people, and humanity's reach exceeded the length of its axe as livestock, first sheep, then cattle, swarmed over the lusher lands. This first boosted burning because transhumant herders fired the mountains, and then destroyed it because the intensive grazing stripped out the grasses that had made routine surface fire possible. Mining drove the system in the Rockies and the Sierra Nevada; agriculture, along the Wasatch. The two practices led to very different forms of settlement.[6]

But everything seemed to boost bad burns. Powell described how he himself had accidentally kindled a campfire against a pine that ignited the trunk, scattered sparks to the woods, and "swept for miles and scores of miles, from day to day." He had personally "witnessed more than a dozen fires" in Colorado of comparable magnitude, and his crews had mapped in Utah a record that testified to "about one-half" of its forests "thus consumed since the occupation of the country by civilized man." Across the region the fires were "so frequent and of such vast proportions" that their smoke stalled survey work. A later 1889 trip by rail found the Dakotas, Montana, Washington, Oregon, and Idaho obscured by a fog of smoke. "Ever a mountain land, and never a mountain in sight."[7]

The climax to the *Arid Lands* report was proposed legislation to reform America's westward movement. Arable farming required irrigation, which demanded public control over watersheds; and since much of the land favored grazing rather than farming, land laws for disposing the public domain should accommodate that fact. Behind those particulars lay a fear of monopolization should private capital be the primary means of allocating natural resources. The counterexample lay before the Powell Survey along the Wasatch. Mormon settlement favored public over private ownership of natural resources. The hierarchy early declared that "there would be no private ownership in the water streams; that wood and timber would be regarded as community property," that the gathering of Zion took precedence over the accumulation of individual wealth. This communitarian ethos, not unlike what he knew from the Paiutes, made Utah—to Powell's mind—an exemplar for how the Arid Region might be made productive, not merely pillaged. The secular state would have to assume the role of Deseret's peculiar theocracy.[8]

What moves the story from another sad anecdote in the disenfranchisement of American Indians and the alliance of state-sponsored conservation with fire exclusion is that Powell subsequently recanted. The reason was his further study of native burning. In 1879, while the *Arid Lands* was being reprinted, he assumed directorship of the Bureau of American Ethnology, established within the Smithsonian Institution, a post he continued to hold concurrently with his later directorship of the USGS. In his explorations he had become fascinated by the West's indigenes, particularly the Paiutes. His attribution of widespread burning had not come from animus but from curiosity and close attention to their circumstances. Removing the natives would indeed remove a primary source of free-ranging fire and thus shield the mountain watersheds from putative ruin.

In 1890, however, while at the height of his powers, he reversed himself and argued that the proposed system of forest reserves, which would be enacted into law the next year, needed to emulate Indian burning if they were to prevent more damaging fires. Worse, he made the case by bursting into a meeting between Secretary of the Interior John Noble and America's two most prominent foresters, Bernhard Fernow and Gifford Pinchot, and tossed his firebomb into their discussion about the hoped-for reserves. Apart from discourtesy, Powell offered a direct challenge to forestry as the appropriate vehicle to manage the reserves. There was, Powell argued, an important distinction between wanton stand-replacing fires and routine surface burning. The former were shredding the western forests and smoking in its settlements. The latter could shield the woods from bad burns.[9]

One reason behind Powell's change of mind was a change in the character of fires, which reflected a shift in land tenure and use. "Before the white man came the natives systematically burned over the forest lands with each recurrent year as one of their great hunting economies. By this process little destruction of timber was accomplished." But when that routine ended, and "civilized men" appeared, the forests rapidly receded, and the fires worsened as "needles, cones, and brush, together with the leaves of grass and shrubs below, accumulate when not burned annually." To remedy that shift Powell proposed some selective grazing, enough to dampen the grasses and shrubs and create trails, not enough to destroy the cover or woods, along with some restored annual burning on the indigenous model. Tame fires could protect against feral fires.[10]

Such a proposal posed a direct challenge to the dogmas of professional foresters, and hence to their authority, and the guild retaliated by dismissing Powell's proposal as "Paiute forestry" (and on the basis of his own testimony accused Powell of vandalism). The next year Congress authorized the creation of forest reserves, and responsibility for fire began shifting from exploring naturalists, geographers, geologists, and agronomists to foresters. In brief, in the heart of the Interior West lay the origins for the argument that the federal government had a legitimate interest in stabilizing and directing the character of western settlement, the question of what agency and professional guild might oversee intervention, the appeal to science as a mechanism to resolve conflicts over the West's natural resources, and the debate between fire fighting and fire lighting as proper policy.

In time Powell's prediction that reservations and maturing settlement would end the fires proved correct. After a wild wave of fires that rode in with the frontier, burned area decreased; ignitions shifted from people to lightning; and for nearly all the century that followed, Utah found itself among the disappeared of America's fire provinces. Colorado, too, after riotous flare-ups associated with its mining boom, faded from the national fire scene. Its principal contribution was a masterful study of fire ecology in lodgepole pine conducted by Frederic Clements in 1910 that slipped into oblivion, overwhelmed by the Big Blowup in the Northern Rockies. Depopulated Nevada became for America's fire narrative what it became for its politics, a Great Rotten Borough. They—the states of the Arid Region, the intermountains—were, collectively, with patchy exceptions, the empty lands of American pyrogeography.[11]

In the 1990s the Interior West began to reemerge into prominence, and by the new millennium it had pounded in fresh claim stakes and geodetic markers in America's geography of fire. Amid the settlement wave of the 19th century, each state made a particular contribution to development. Colorado was the font for western water law. Nevada had birthed the notorious mining acts of 1866 and 1872. And Utah, thanks to the Powell Survey, argued for state intervention to prevent monopolies of natural resources. Now, amid the late 20th-century resettlement of formerly rural lands that is spilling into the 21st century, they again promised to contribute to the national narrative.

Their economies had firmly shifted from commodities based on rock, water, grass, and timber to services based on tourism, gambling, recreation, and exurban sprawl. The front ranges had become front lines in the sad, lethal story of spark and sprawl. Colorado rivaled California for fire fatalities and for losses of high-end homes. Nevada's sage grouse threatened to do for the Great Basin what the northern spotted owl had done for the Pacific Northwest. Utah's high plateaus brought a national narrative full circle, as prototypical fire mapping returned to its place of origin. Instead of making the desert bloom the state was watching its high forests burn, along with its scruffy woodlands and battered grasses. The empty hole in the center of the country had filled with combustibles and was reburning and reclaiming national attention as it did.

The once anomalous arid lands hinted at the harbingers of a new national normal and perhaps a new, unexpected national exemplar. That Utah would, first among American states, enact the precepts of the National Cohesive Strategy into law would seem to come with the force of an alien visitation. Or it would, had Utah not already known that role 140 years earlier.

NEVADA

FROM ROTTEN BOROUGH TO BURNING MAN

IN 1859, 10 YEARS after forty-niners had dashed across the country in search of California gold, restless prospecting found the Comstock Lode. Here—at Mount Davidson, in the first range east of the Sierra Nevada, between Reno and Carson City—was a mountain of bullion that inspired the country's first great silver rush. Miners spilled over from the Sierra crest. Mormons had meanwhile established small settlements along the Humboldt River and in Reno. The population swelled sufficiently to warrant statehood. The land lay in the territory of Utah, however, and not wishing to grant control to the Mormons' shadow empire, Deseret, and hoping to boost the reelection of Abraham Lincoln in 1864, a new state was created, Nevada.[1]

The Comstock Lode played out by 1874, though other mines were found in the exposed slopes of the Basin Range. So while the population of Virginia City crashed, the state's population remained more or less constant, floating from strike to strike. The 1900 census gave the same figure as the 1870 (42,000). Ranchers moved in; the population of sheep, and then cattle, swelled. If mining might be reckoned as a series of stampedes, ranching might be characterized as a series of rushes. Then the winter of 1889–90 destroyed the herds. The upshot was a Great Basin filled with the biotic equivalent of slag heaps and holes. Virginia City's population in 1880 was 10,719; in 1980, 600. Writing in the early 1940s Wallace Stegner noted that "even in terms of towns, the dead in the Great Basin outnumber the living."[2]

Nevada began as a bastard child of California, and continued to play silver to California's gold. Its most populous towns were those hardwired to California—Reno to Northern California, and in the post–World War II era Las Vegas to Southern California. Its small, unstable population, largely committed to two industries, both vulnerable to railroads, rendered it pliable to big money and corruption notorious even by the standards of the Gilded Age. It became the "great rotten borough" of American politics, and once it found a new source of bullion in divorce, gambling, and legalized prostitution, perhaps of the American economy as well and even the national culture. Whether the postwar boom has caused it to change its stripes, or has only made them larger and more garish, the future will have to look back and judge.

The past's judgment on the land has almost uniformly been one of indifference, if not disdain. The state's preponderant population, miners, rarely saw beyond the "color" in the veins of outcrops. Others just passed by. Mark Twain set the tone when he dismissed the sagebrush steppe as "the fag-end of vegetable creation," and in *Roughing It* called it a "very fair fuel," but a "distinguished failure" as a plant. "Nothing can abide the taste of it but the jackass and his illegitimate child the mule." His successors left the Great Basin "one of the least novelized, least painted, and least eulogized of American landscapes." Beyond bullion there was almost no cultural engagement, no national institutions committed to the place, no intellectual values in what it held. It was a giant sinkhole in the path of American progress.[3]

Its fires followed the same rhythms and paces. The barrenness that made prospecting easy also kept combustibles for burning small. The population of natural fires, like its indigenous population, was sparse. There were fire rushes as miners flooded in, then fire ghost towns after they had stripped the countryside bare before leaving. Until the U.S. Forest Service arrived in 1905, there were few formal fire agencies. There were none for the vast expanses of the steppes. The Great Basin—Nevada outside its eastern Sierra slopes—was empty of significant fires, fire institutions, and fire culture.

Fire requires fuel, which requires life, which requires water. The Great Basin, however, is a desert. Its streams end in saline lakes or salt flats or just sink in the sands. (Its principle watercourse, the Humboldt, which also served as thoroughfare for westering forty-niners and pioneers, was long known as the Inconstant River.) Its primary rains for grasses come in the fall, outside the usual regime. Its spotty human settlements cluster around its perimeter, where rivers debouch from the mountains; the exceptions are mining towns, which rely on groundwaters, and Las Vegas, which is a city-state with its own idiosyncratic imperative. The human economy clusters around the urban centers that peg down the edges of the Basin, like stakes around a drying hide. Boise and Salt Lake City lie outside Nevada. That leaves Carson City-Reno and Las Vegas, which bind it to California, like hinges on a door.

Until it reached the Rockies, America's westering had proceeded like a slow debris wave of flame, as people used fire to convert the encountered landscapes into fields, pastures, and towns. That wave broke when aridity became a defining attribute and fuels lost continuity. Agricultural landscapes no longer provided an interstitial medium, a connective tissue, to hold the pieces together. The Great Basin had many ways in, few out, and most newcomers wanted to pass through as quickly as possible.[4]

Settlement hopscotched around according to mines and pockets of irrigable land and native pasture. Anthropogenic fire became patchy, a scatter of spot fires rather than a creeping front of flame. In America's humid east, ranching encouraged more fire as herders burned to encourage pasture. In the arid west, ranching broke the old fire regimes by cropping off the necessary combustibles. Continuity didn't return until invasive plants, notably cheatgrass, claimed the landscapes vacated by the overgrazed perennial grasses and rewrote the fire regimes of sage steppes and desert playa.

In C. S. Sargent's cartography of American fire for the 1880 census, Nevada was blank save for the mining regions around the Comstock (up to Lake Tahoe). The only fires that concerned anyone were those that incinerated the ramshackle wooden towns. Virginia City burned routinely, as most such self-combusting places did; the October 26, 1875, fire leveled the town and even burned 400 feet into the main shaft of the Ophir Mine. In the backcountry lightning kindled fires on the mountains and on the sage steppes, prospectors burned off hillsides covered with vegetation in order to expose outcrops, and herders fired the summer pastures when they departed in the fall. There were the usual abandoned campfires, accidental ignitions, and stray incendiaries. There was no organized system for reporting fires, fighting fires, or regulating fire usage.

That began to change in 1905, when the Forest Service assumed responsibility for the forest reserves on the higher of Nevada's ranges. In 1934 the Taylor Grazing Act closed the public domain to further patents and sought to bring some order to the laissez-faire livestock industry. Its successor, the Bureau of Land Management, did not become serious about fire protection until the 1964 Elko fires. By then cheatgrass and its invasive cousins were remaking the fundamentals of fire in the Basin. The State of Nevada did not become serious about fire until flames began to threaten its two urban clusters, running into Reno and Carson City and even rampaging over Charleston Peak outside Las Vegas. The catalyst for reform, however, was the threat to list the greater sage grouse as endangered. The only immediate fix to the destruction of its habitat was aggressive fire suppression.

The Great Basin became as eccentric for its fire history as for its settlement patterns. It hosted the quirky, the disjunctive, the fires that seemed to belong nowhere else or had nowhere else to go. The Burning Man carnival, the Nevada Test Site, the fires that flared up like summer fireworks on isolated sky islands and then went dark, even the ravenous cold-desert steppe fires that ripped through cheatgrass—these added up to a cabinet of curiosities, not a coherent panorama. If not exactly the fire equivalent of the Las Vegas Strip, they had no reason to belong together beyond the accident of political geography that gathered them into a state called Nevada. They did not fit tidily into prevailing norms or scientific paradigms or national narratives.

Until recent times it's hard to find a Nevadan culture of fire. The population only concentrated where bullion did, whether in placers, quartz veins, or casinos; and a culture based on minerals is not one inclined to burn landscapes except accidentally or in anger. The Burning Man ceremony sits within a salt playa at Black Rock, sited there precisely because it cannot propagate except in the imagination. The nukes detonated over Frenchman Flat were a geopolitical flash fire, for which the term "fire regime" hardly applies except by linguistic acrobatics.

So fire does not speak, even in the background. Mark Twain in *Roughing It* penned one of the two most famous examples of writers accidentally starting a fire when he managed to set the borders of Lake Tahoe aflame (the other is Henry Thoreau, who clumsily set the Concord woods on fire). When the mining era faded into ranching, Walter Van Tilburg Clark's novels of ranch life had no fires but do convey the dark, self-destructive side of settlement.

> "Meaning if one Comstock's used up, there'll be another?" Arthur asked softly. "A bigger and better one?"
>
> "You goddam right there will," Curt said, "for the guy that knows when he sees it . . ."
>
> "But for everybody? No. That was a kind of dream too, a big, fat one, and it's over. We've gone from ocean to ocean, Curt, burning and butchering and cutting down and plowing under and digging out, and now we're at the end of it. Virginia City's where the fat dream winked out. Now we turn back."

In the end, still fighting the natural scene, Curt destroys himself.[5]

In the postwar era—or as some would have, a postindustrial and postmodern era—the dream of another Comstock returned in the form of Las Vegas. Vegas now dominates Nevada culture, in all media, as it does in population, but the 1980 MGM Grand fire that killed 85 and injured 700 is far removed from the saga of landscape fire, and no one has managed to turn fire into a plot pivot. Only when the threatened listing of the sage grouse put the national fire establishment in a headlock has Great Basin fire claimed part of the greater stage. A hundred acres burned outside Santa Barbara or Hollywood is national news. A hundred thousand burned outside Winnemucca barely merits a byline.

Where it figures in the national narrative is its role in bolstering the BLM as a fire agency, which went a long way to elevating the BLM into prominence as a land management agency. Outside the 1964 Elko outbreak, which was a catalyst for transferring BLM fire operations from interior Alaska to the Interior West, it's hard to find burns, however vast, that moved fire seismographs elsewhere.

Battle Born. It's Nevada's state nickname, along with sagebrush state and silver state. The silver still flows, though through slot machines. The sagebrush still covers most of the state, though it's receding in the face of cheatgrass-powered fires. And Battle Born persists as part of a new civil war in which the American political system has become increasingly polarized and paralyzed, in this case over environmental controversies. Despite its transient population, Nevada the state was created hastily and admitted to advance a political agenda. And despite its peculiar values at risk, Nevada the fire province now finds itself again enlisted in a national contest to decide the future directions, or even the possibilities for, management of America's wildlands. It may not be coincidence that the Forest Service director for Fire and Aviation Management, Shawna Legarza, appointed in 2016, grew up on a Nevada ranch and began her career with a BLM engine crew outside Elko.[6]

The sage grouse promises to make Nevada relevant to national fire establishment in ways the northern spotted owl did not. For the Pacific Northwest the story was about staying the axe, a social and political decision about human behavior, with the owl as mediator, token, and cipher. For the Great Basin the story is about staying fire, which is not something solely in the hands of people but a negotiation between people and nature, for which the sage grouse may be a chip in the ante but cheatgrass is the trump card.

THE OTHER SIDE OF THE MOUNTAIN

Washoe WUI

THE SCENE IS both common and eccentric.

The familiar is a swathe of landscape along the Sierra Front that resembles sprawling urban centers everywhere in the West. Like a giant mall designed to carry traffic between two anchor stores, the Sierra Front spreads between Reno and Carson City, and probably to Garnerville. It also looks like urbanization elsewhere in the region, all of which cluster along the fringes of the Great Basin. And like Nevada's other urban complex, Las Vegas, this one began as a corridor to California and remains tethered to it.

But the oddities matter. Here the shoulder of the Sierra Nevada broadens into rough terraces that step down from the granite crest to the saline lakes of the Basin Range. Reno and Carson City sit within valleys that echo Lake Tahoe, though broader and lower; the scene less concentrated, the lakes shallower, the vegetation drier. The Sierra Front suffers far less aggressive urbanization than Vegas but is far more vulnerable to fire along its edges. The Mojave Desert surrounds Las Vegas, even if the city's exurbs are testing the fire-prone slopes of Mount Charleston. The eastern slope of the Sierra Nevada engulfs its creeping sprawl with a medley of conifers, sagebrush, and cheatgrass that burns briskly and occasionally spills into town. The fireshed more resembles a dry Tahoe than it does Boise or Salt Lake City.

Historically, too, the Sierra Front—call it the Washoe WUI—knew a fire scene far more boisterous than its Great Basin cognates. The mining boom that washed over the Sierra crest and spread from regional entrepôts like Reno led to a riot of burning. Sargent's 1880 map of forest fires identifies the zone, extending through the mining regions of the Mother Lode country into San Francisco, as a major hot spot in the American West.

There was probably not much left after the miners overran the landscape. They cut and burned timber, they used sagebrush for fuel, they ran livestock over whatever cattle and sheep could feed on. Wood-framed Virginia City burned much as early San Francisco did, and as often. Then the ore played out, the population drifted away, and the land began a slow rehabilitation. Silver being where you found it, that left the siting of forest reserves where mining claims weren't, which meant that national forests along the Sierra slope were patchy. After the Taylor Grazing Act, the unpatented public domain eventually fell to the BLM. So the Forest Service oversaw the public lands to the west, and the BLM those to the east, forming rude levees between which the rural countryside rested, and through which urbanization later flowed. By the late 20th century some of these villages would be massing into towns.

Each of the lands had its own characteristic fire problem. Each had its own system for fire protection. There was one arrangement for wildlands, one for rural lands, and another for cities. The scene resembled Southern California with its mosaic of jealous jurisdictions, its sprawl crowding against mountains, and Donner Pass, almost as vital a transportation and utility corridor for the North Coast as Cajon Pass is for the South. But until the lands recovered, or renewed in fire-prone ways, and until the rural scene reincarnated as suburban or exurban, until, that is, the fires could spread more vigorously and push beyond the countryside into the city, there was little incentive to do more.

The smart money was that in fire protection, as in other matters, the Sierra Front would follow California's lead. In the postwar era it did, and then it didn't. The Front lies within the fire shadow of the Sierras. The crest intercepts most of the rain and snow. So, too, it filtered fire institutions. It tells the less celebrated story of the other side of the mountain.[1]

A working approximation of fire behavior need only know the place and timing of ignition and the character of the winds. This certainly holds for the Sierra Front.

The Front north to the Modoc Plateaus is dense with fires. Lightning is plentiful, and as storms move along it they appear to veer either north or east, with particularly hot cells kindling fire busts. But almost as common outside urban areas are human causes from ATVs, shooting, camping, fireworks, and recreation generally. Unlike lightning these starts are not seasonal but constant, and unlike wildland starts they crowd around areas people live in or frequent. Most starts (over 90 percent) occur at lower elevations, on private land. That puts fire in the most threatened landscapes. What happens next depends on the wind.

The prevailing southwesterlies tend to drive big fires cross-grain to the terrain. The most explosive conditions come from passing cold fronts, which quicken that southwesterly flow before switching to northwesterly winds that dry as they rush over the Sierra crest. The dramatic terrain causes plenty of local eddies and eccentricities; the pass created by the Truckee River over the crest and down toward Reno, for example, creates its own flume. And then there is the summer's 2 p.m. downburst, the famed "Washoe Zephyr." Let Mark Twain describe it.

> This was all we saw that day, for it was two o'clock, now, and according to custom the daily "Washoe Zephyr" set in; a soaring dust-drift about the size of the United States set up edgewise came with it, and the capital of Nevada Territory disappeared from view. Still, there were sights to be seen which were not wholly uninteresting to newcomers; for the vast dust-cloud was thickly freckled with things strange to the upper air—things living and dead, that flitted hither and thither, going and coming, appearing and disappearing among the rolling billows of dust—hats, chickens, and parasols sailing in the remote heavens; blankets, tin signs, sage-brush, and shingles a shade lower; door-mats and buffalo-robes lower still; shovels and coal-scuttles on the next grade; glass doors, cats, and little children on the next; disrupted lumber yards, light buggies, and wheelbarrows on the next; and down only thirty or forty feet above ground was a scurrying storm of emigrating roofs and vacant lots.
>
> It was something to see that much. I could have seen more, if I could have kept the dust out of my eyes.
>
> But, seriously, a Washoe wind is by no means a trifling matter. It blows flimsy houses down, lifts shingle roofs occasionally, rolls up tin ones like sheet music, now and then blows a stage-coach over and spills the passengers;

> and tradition says the reason there are so many bald people there is, that the wind blows the hair off their heads while they are looking skyward after their hats.[2]

And it can stir a fire into a little excitement.

The fires obey the winds. The typical problem wind lasts from one to three days. Either you catch a start before the wind can bend it or you wait for the wind to stop.

The Washoe WUI is the defining feature of the Sierra Front's fire scene. The risk runs with the sprawl, and those lands are private. National forests drape the eastern slopes of the Sierra, but patchily. Public lands under the Bureau of Land Management flank the eastern fringe of an interface that near towns has the character of an intermix. Not only does this de facto zoning stiffen the edges of the Washoe WUI, but it makes fire protection possible. The feds provide muscle, money, and organization that boost local capacity. They don't have much land in the WUI proper, but they contribute some fires and high-volume, high-intensity firefighting such as aircraft. Anything that diverts their attention affects their overall ability to participate. The BLM in particular has another landscape almost as compelling, the sage steppe habitat of the greater sage grouse. When fire busts occur, they are as likely to happen in one habitat as the other.

Along the Front proper, the problem fires burn as brush fires, with cheatgrass acting as an accelerant. On its western flanks, the cheat and sage grade into conifers, all recovered from the flush times of the Comstock, and all matured under climatic conditions that no longer apply. On its eastern flank, in the six-million-acre domain that comprises the BLM's Carson City District, the problem fires burn in cheatgrass, rising and falling with the unsteady rains that grow or starve annually. No cheat, no fires. Dense cheat, intense fires. The economy of fire resembles Nevada's classic mining economy as it swings in and out of boom and bust. (Since 1999 the Great Basin has oscillated on a three-year cycle—three years of big burning, three years of small.) The rhythm applies to fire's management as well, since a recruit might spend two or three seasons without experiencing a serious fire. Along the WUI an engine can be at the scene in minutes. Among the stage steppe it can take hours, even the better part of a day. And aircraft alone cannot substitute for ground crews.

Behind each agency stretch larger firescapes. California—Forest Service Region 5—looms over the Sierra Front as the crestline does the east-facing slopes. The gravitational pull is immense. California is a rapid reservoir of fire resources, one that can be mobilized directly without going through larger regional or national dispatch centers, but those crews, engines, and aircraft come at the cost of accepting California's norms; when they meet, fire officers must wear brass on their lapels. The Sierra Front falls within Region 4, but is often mockingly referred to as Region 4.5. Like the Washoe Zephyr, the local institutional winds blow from California.

The BLM, too, has a second sun around which it must orbit. When Secretarial Order 3336 mandated a strategy of protection for sage grouse habitat, it forced the agency to fight on two fronts. Both firesheds are expanding. Both demand full suppression. Both argue for an initial attack based on overwhelming force—not as overpowering as that typically dispatched in Southern California, but sufficient to halt spread before the winds can sail flames through the countryside. The costs—fiscal, ecological—can be high.

The Sierra Front must also factor in a managerial cost. The paradox is that a rapid, multiactor response can be as daunting as the fire. Fire suppression resources are generally adequate, and more can be ordered quickly from California down I-80 over Donner Pass. The proliferating number of fire agencies along the Front can each send engines. The concern is not with inadequate response but with the complexity of organizing that response into coherence. By size typical fires should favor Type III overhead; by complexity they slide into Type II or even Type I. When smoke rises along the fringe, half a dozen or more fire jurisdictions may respond. The big call-outs can jam up roads, clog effective action, overwhelm overhead. Engines have burned on the roads. People have been threatened. In 1983 miscommunication during the Mound House fire led to a firefighter fatality.

Like the WUI, fire institutions "just growed" without planning. A threat in one part of the Front was a threat to all: they all responded, not with zoning or regional planning boards, but with visible public-safety measures in the form of engines. There was too little preventive action, too much emergency reaction. To help one another they had to get out of each other's way in the field and at each other's side in committee rooms. The scene was not unlike that which prompted Southern California to create Firescope, which evolved into the incident command system. The Front followed that example, but with local flavoring.

In 1982 the fire community came together as the Sierra Front Wildfire Cooperators. The stimulus was the immediate embarrassment of ineffective over-response and the unfolding crisis of urban sprawl. In conception it was an all-hands approach to improve the fire scene. It sought to make the Front less fire prone and fire agencies more coherent when working together. Arguing before the Nevada legislature to have fire included in master plans was a gamble, but there was no excuse for engines not to use the incident command system, not to have the same tactical frequencies, not to pile up in staging areas because no one was empowered to say who should go where. By the mid-1980s the Sierra Front, like most of the West, was experiencing an inflection into larger, nastier fires.

The Front was a bold idea, and a slow process; the interagency idea as a default setting was still years away. There was a major revision to the agreement in 1996, and another in 2010. By then there were 15 entities in the coalition. The University of Nevada cooperative extension program promoted a Living with Fire Program, and Washoe County organized a Fire Adapted Communities Project. With assistance from the National Fire Plan, the cooperators developed a Carson Range fuel reduction and wildfire prevention strategy. The Sierra Front was looking like a lot of at-risk fire zones, but more serious, and with a longer history of cooperation.

By the early 21st century, between the Washoe WUI and the threatened sage grouse, the Sierra Front was leaping from sideshow to frontlines in the nation's pyrogeography. The best guess was that it would evolve as similar firescapes did elsewhere. In particular, it would look more and more like California.

For decades it did. The keystone would be the State of Nevada, which had little direct holdings in the Front but had indirect jurisdiction over much that happened.

Nevada created a state fire program in a scenario that was typical for the West but slower to evolve. In 1931 the state began negotiating with the U.S. Forest Service under the Clarke-McNary Act to create an embryonic state fire protection system. Six years later a statute authorized fire protection districts. In 1945 the Nevada legislature established the office of state forester fire warden to oversee some 8.7 million acres. Steadily, powers and responsibilities were amended, and mostly expanded into something akin to a state forester. In 1957 a Department of Conservation and Natural Resources was authorized, which oversaw a Division of Forestry, which absorbed the state forester fire warden. In 1959 a program of inmate honor camps was created. A typical series of reorganizations followed. The federal 1978 Cooperative Forestry Assistance Act replaced and expanded the Clarke-McNary Program. The Nevada Division of Forestry (NDF) had responsibilities for fire protection over state and private lands. It looked as though the Nevada Division of Forestry might follow the example of the California Division of Forestry and become the state's primary fire agency.[3]

Instead, the program devolved. Beginning in 1985, and continuing for 30 years, the NDF transferred responsibility to the counties and fire protection districts, while it provided assistance, especially financial, with the big burns. A formula evolved in which the values at risk and historic fire loads set the base funding each county would furnish. Above that, the state could assist with engines, crews, and overhead, and of course with agreements with the federal agencies that granted access to their resources; and NDF would pay for the cost of large fires once the county satisfied its base obligation. Given Nevada's geography this made sense. Some 80 percent of the land was federal, and 90 percent of the population was crowded into two urban clusters. Locals needed help. The NDF, however, could not cover the state as CalFire could try to do for California. Still, it became more involved in direct action. A master fire agreement codified the arrangements. The 1990s boomed.

Then the crisis hit. The Great Recession slammed Nevada hard. Its immense housing bubble burst. The state treasury nearly went bust. So just as the fires were getting bigger and the risks higher, the capacity to fight fire threatened to collapse. The NDF shrank. Funding fell, staff and resources were laid off, and the scope of its mission challenged. The governor directed every state agency to justify every task by pointing to an authorizing legislative statute. Devolving slid into dismantling.

California found ways to keep up its extraordinary fire protection system during the recession. The other side of the mountain, Nevada, did not. In 2012 Bob Roper, formerly fire chief for Ventura County, California, was appointed forester fire warden. It seemed clear that NDF would have to survive by collaboration, and the National Cohesive Strategy suggested what that might look like and what a suitable mission statement might embrace. NDF contracted for a strategic plan to be written that would move from the vision of the National Fire Plan to that of the National Cohesive Strategy. The plan's opening article stipulates that NDF will "adopt the tenets of the National Cohesive Strategy and adapt it to Nevada," what would be a Nevada Cohesive Strategy.[4]

Still very much a work in progress, the plan shifted NDF from direct actor to supporter, facilitator, and partner. It could furnish some labor and equipment, but mostly it advised and coordinated, and connected local needs to regional and national resources. Then, with a new director and its mission still inchoate, the Little Valley prescribed fire, set on state lands under restoration west of Carson City, escaped on October 14, burned 22 houses, and kindled a political firestorm. The strategic plan remains in edit. The NDF is still feeling its way to a future.

Little in fire management is not connected with everything else. The Humboldt-Toiyabe National Forest must reconcile the Sierra Front with the national fire borrowing crisis. The BLM must balance its commitment to front country WUI with its commitment to backcountry sage grouse habitat. The NDF must juggle prescribed burns in Little Valley with protection assistance in Elko, Winnemucca, and Caliente. No one can do it alone, yet the limitations and distractions of each affect how the whole functions. The cooperators weaken or strengthen with the ebb and flow of subprime frauds, fickle publics, and the Washoe Zephyr of personalities. The pact jostles into new adjustments as each partner strides or stumbles.

The land, too, is reorganizing, as it has since the Comstock collapsed. But its assembling parts have seemed to move in only one direction. As the NDF's new administrator, Joe Freeland, observes, along the Front you see "either a fire scar or a problem." Most likely you see both.

A SINK FOR EXOTICS

FOR MUCH OF its post-Pleistocene history, the Great Basin has been short on natives and long on newcomers, most of whom came, ran wild, and left. The recession of its great pluvial lakes left the Basin as emptied of an enduring biota as of waters. What the Pleistocene had filled the Holocene drained. What blew in mostly blew out. That left a lot of emptiness in a big place. If the Great Basin were a state, its size would rank third, just behind Texas. There were geographic gaps unplugged, ecological niches unoccupied, a biotic basin unfilled. There were apparent voids ready to receive Burning Man, weapons tests, nuclear waste, even the MX missile scheme that was intended to absorb most of the USSR's nuclear warheads. Migrations were still in motion. The natives were often poorly equipped to resist species forged in hardier fire regimes elsewhere. The enduring issue was not those who came and then left but those who came and stayed. There were plenty of exotics ready to enter. There were, in the end, too many of the wrong kind who stayed.

It's a harsh land—a desert, as the locals like to remind visitors; hot in the south, cold in the north, but almost everywhere arid. Nevada's invasives are less numerous than, say, Florida's, but they and their damages are more visible. Nothing remains hidden for long in a desert.

INVASIVES: A SCORECARD

The big shock was European contact and a riotous American settlement. Native populations plummeted, and whatever effects they had on the indigenous biota, including fire, shrank with them. As so often in its *longue durée* the Basin emptied, then began refilling, though this time at a rate perhaps rivaled only by the catastrophic draining of 1,200 cubic miles of Lake Bonneville through Red Rock Pass 15,000 years ago. Exotic plants and animals entered, then often exploded through the ecological voids. Miners stripped hillsides of trees for timber and fuel. Ranchers unleashed cattle and sheep, and often kept them grazing continuously until the perennial grasses died off, and then burned sage steppes to produce more. Farmers (and most ranchers also farmed for winter fodder) introduced weeds along with wheat, and migrating harvesters scattered them further. The majority of the newcomers burned, at least in patches. Later, a sentimental attachment to wild horses replaced managed livestock, whose numbers were regulated, with feral ungulates, which propagated unchecked. The Great Basin held the dream team of American invasives.

But the major catalyst was the railroad, which connected ranches, farms, mines, and townsfolk with markets. Without rail there would have been no economical way to dig ore, stock ranges, or sell goods. The larger fire economy of the Great Basin shifted—no, shifted is too mild a term for what must have resembled an overturning iceberg—from burning living landscapes to burning lithic ones. The fundamentals no longer lay with climate and migrating species, and whatever fires lightning and hunter-foragers kindled, but had to cope with another combustion regime, one with global reach. It was like adding a second sun.

The compounding havoc seems obvious only in retrospect. At the time there was only this or that observation, an astute or throwaway comment or anecdote, often by travelers themselves new to the country, for whom the disturbed landscape evident when they arrived seemed natural and served as a baseline for whatever disruptions they recently noted. Later, as the totality of change became undeniable, professionals—academics, research scientists, naturalists, range managers—abstracted, simplified, and reduced the outcome of thousands of individual choices and chance events into ecological forces, climatic stresses, wholesale movements and migrations of people, plants, and fauna.

Yet the reality was a matrix of particulars. The farmer who welcomed new, high-value grain from California, one spiked unknowingly with seeds from central Eurasia. The harvester, one of many mechanical migrants, going from field to field, carrying seed in the axles of his thresher. The small rancher who kept his cattle on the same field year-round because he had no choice. The herders who lit fires to promote grasses instead of sagebrush and to green up mountain pasture. The railroad, the cattle tracks, the dirt roads—all became pathways for ecological pathogens. The soils and rainfall regime were ideal for exotics from central Asia but they needed vectors to transport vagabonds from the goosefoot and mustard families, and the ecological carpetbaggers that comprised the clan of bromes.

Those small acts added up. A plague only requires a few points of infection in a receptive population. In the Great Basin preadapted plants met a prearranged landscape. The upshot was not the result of simple malice or reckless risk taking or pervasive greed. It followed from the compounding effect of individual coin flips that kept coming up heads until everything changed and the old order was no longer probable. Still, the scale of the transformation was breathtaking. The sage steppe survived in refugia distant from roads and ranching.

The old order of ecological successions crumbled: disturbance led to more disturbance. What may be most remarkable is that the upheavals didn't end with the new aliens. The disruption became constant, and a new pattern of succession arose in which, in more or less predictable sequence, the invasives themselves were invaded. First, the gooseberry family, then the mustard family, then the tribe of *Bromus*: Russian thistle, Russian wire thistle, tumble mustard, annual kochia, red-stem fillaree, red brome, cheatgrass. The implication was that today's invasives would become tomorrow's naturalized indigenes to be overrun by tomorrow's exotics. The cheatgrass infestation might not be a one-off replacement of perennial grasses and sagebrush but a phase in a long history of once and future migrations. The historical logic of this observation is that the control of cheatgrass may depend on the arrival of an even more durable invasive.

CHEATGRASS

The rap sheet on cheatgrass (*Bromus tectorum*) is long and growing. It's not even the case that cheatgrass is a recidivist offender because it was never put away.

It's an old species. It was listed by Linnaeus in 1753 and was known as a common feature of grass roofs (hence its name). Today it can be found in every American state, though in most it's a minor face in the crowd. In the Great Basin it found conditions that allowed it to spread like an internet meme. It first appeared along rails and roads and the edges of hay fields around 1900–1910. By the 1920s it had propagated so widely it had become the wanted poster of western weeds, notorious enough that even Aldo Leopold could imagine it as a "notable instance" of an exotic species whose "spread was so rapid as to escape recording; one simply woke up one fine spring to find the range dominated by a new weed." Its propagation defined a century. Cheatgrass and rangelands in the 20th-century Great Basin are, as its principle chroniclers James Young and Charlie Clements put it, "synonymous."[1]

Cheatgrass, they conclude, "transformed Great Basin environments and then evolved to exploit the changes." When abusive, continuous grazing broke native grasses, all perennials, cheatgrass moved in, and then outcompeted them for critical early-spring moisture. It next destroyed the shrubs by encouraging fires, which in a poster-child model of a positive feedback loop, furnished ideal conditions for reseeding cheatgrass. That eventually impoverished native wildlife and birds, save for granivorous rodents who came to prefer its seeds. As the dominant survivor, cheatgrass became, by necessity, a critical surrogate spring feed for such introduced fauna as livestock and game birds like the chukar partridge. The fires have become more immense, more damaging, and vastly more costly to combat, and they leave in their wake a landscape fit best for more cheatgrass. Only wholesale intervention might slow it down, but unless done thoroughly remedial measures promise only to aggravate the scene. (While reductionism is the essence of the scientific method, it's a formula for failure in resource management.) "Above all else," Young and Clements note, cheatgrass represents "a stage in transition toward an environment dominated by exotic weeds growing on eroded landscapes." It's a biotic version of a race to the bottom. Call it ecology's *Blade Runner* scenario.[2]

The plague might have passed as a silent and measured infestation in an empty region were it not linked with fire. Cheatgrass-dominated landscapes claim 6 percent of the Great Basin, but cheatgrass itself is nearly everywhere, and increasingly it defines the fire scene. The presence of cheatgrass inflates all the combustion properties. Fires are more easily ignited and more readily spread. Fires burn four times more frequently in cheatgrass than in native vegetation, burn more than double the area, and burn 24 percent of the area of the 50 largest wildfires. What had burned in patches now sweeps over whole landscapes. In 1973 the Hallelujah Junction fire, considered a monster, burned 35,000 acres, which put it nearly off the charts. In 1999 more than a million acres burned in 10 days. In 2000, 700,000 acres burned; in 2001, 654,000; 2005, 1.7 million; 2006, 1.3 million; in 2007, 890,000. In 2016 the Hot Pot fire north of Winnemucca burned 122,000.[3]

The exact figures vary by local conditions: the Great Basin is a big place with lots of nooks and niches. Nevada is the seventh largest state, almost as big as New England and New York together. Nye County is the size of New Hampshire and Vermont. Humboldt County is the size of Massachusetts. But pure stands of cheatgrass are only the most visible manifestation of the reformation. By mingling with sagebrush, pinyon-juniper, and saltbrush, cheatgrass has boosted their burning as well, like adding lighter fluid to charcoal briquets. Since the postburn setting favors cheatgrass over indigenous rivals, the cycle not merely repeats, but enlarges.

Those fires connected cheatgrass with national interests. Previously, fire in the Great Basin had been an occasional event of local significance but not a management problem, and certainly not a national concern. Its social consequences were patchy. Now it is *the* problem: the environmental axle on which everything else in the region turns, and it has perturbed even the national fire establishment. Over the past century it has acted on American society as it has the biota, and often with positive feedback. It sparked a science to study it. It nurtured fire protection agencies to cope with its propagation. And recently it has birthed a crisis in environmental management by its threat to the sage grouse. Its history is intimately intertwined with humanity's. It has infiltrated ideas and institutions as fully as it has big sagebrush and desert chenopods. Range science and weeds, cheatgrass and the BLM fire program—all evolved in a kind of weird co-dependency.

Cheatgrass has proved a wily, flexible species, ready to adapt phenotypically and even genetically. It can be to ecosystems what malaria and HIV are for humans or Amazon for retailers. In many ways it is, like rats, an ideal camp follower for a disturbance-ridden presence like humans. Cheatgrass flourishes amid the disruptions, which it further disrupts. It was a boom-and-bust species for which consistent management was impossible. It thrived amid the overturning of the old regime, and then it thrived amid their attempts to contain it or to restore that imagined former regime. If range managers perturbed cheatgrass-laden sites by burning, continuous grazing, plowing, or poisoning, cheatgrass avidly seized the exposed land. If they lightened grazing, cheatgrass propagated. If they did nothing, cheatgrass still spread.

As many literary-minded naturalists have observed, people pass over lands like a flame, transforming all they touch. Humanity was made for a weed like cheatgrass, and cheatgrass, it would seem, for them. The only solution was to out-cheat the cheat.

CULTURAL INVASIVES

The emptying of the Great Basin by American settlement had emptied it of long-acquired human understanding as much as it had with perennial grasses. What happened to species also happened with traditional ecological knowledge: it seeped away as part of the larger void, and the resulting vacuum filled from the outside. In particular it filled with folklore acquired in places like Texas, and it became prime habitat for an aggressive form of intellectual inquiry, modern science, that, like cheatgrass, had emigrated from western Europe. Clementsian ecology, European forestry, and newly emerged hybrids like range science did for Great Basin culture what invasives did for sage steppe ecosystems. In fact, formal understanding came to resemble the bromes it studied. Rangeland science underwent its own cycles of adaptation as it selected among ideas either substantiated or discredited, as it found competing parties to celebrate or denounce its arrival, and as it created conditions that promoted its own spread at the expense of its indigenous predecessor.

This was a science that first matured in the 1930s, sparked by crisis. Its epicenter was the Great Plains, where its principal proponents came from. Cheatgrass did for the Great Basin what blizzards of dust did for the Great Plains. In both cases abusive land use interacted with climate to raise vast clouds of dust or smoke. In both cases, there were critics who denied any crisis, who insisted that cycles of drought, dust, and fire were normal, who argued that sagebrush had always expanded and contracted, who found pinyon-juniper as troubling as brome. Usually they were less skeptical of the science than of the politics of state-sponsored conservation that the science might provoke (not unlike some climate change deniers today). The Great Plains and the Great Basin underwent a kind of parallel processing by an emergent science that was defining itself even as it was trying to define what, in these places, the problem was it was intended to address.

Cheatgrass and western rangeland science would shape each other. Certainly cheatgrass, along with some other exotics, stimulated the funding behind range science. If it was a problem, then ranchers and agencies would look to science to correct it, and a positivist science would propose solutions. In 1912 the U.S. Forest Service established the Great Basin Experiment Station in the Wasatch Plateau to support research. In 1958 the Agricultural Research Service consolidated pieces of its programs into a Great Basin Rangelands Research program at Reno. The challenge was first to understand the scene and then to correct, deliberately, with informed research and enlightened bureaucracies, what ignorance and indifference had trashed.[4]

As early as 1914 H. T. Shantz identified a cheatgrass-fire cycle, but the foundational paper came from G. D. Pickford in 1932 in which the mating of "continued heavy grazing" with "promiscuous burning" birthed the bastard grasslands that weeds infested and runoff eroded. The critical marker was the appearance of cheatgrass. The genealogy of scientific research follows from this patriarchal study. Those weeds, however, challenged the paradigms of the emerging science; they proved as disruptive to theory as to landscapes. As much as ranchers, rangeland scientists would have to accommodate the new reality cheatgrass created.[5]

Like most scientists, range researchers believed they needed an equilibrium point from which to assess, or else they were only measuring change upon change without end, which made the premise (and promise) of a true science, prediction, impossible. Without that baseline, range ecology would behave less like a science than like a history. Clementsian theory located that stable landscape either in a precontact past or in a putative climax stage to which every biotic community trended. A community would continue to move—that is, shuffle through a predictable sequence of reorganizations—until it reached that resting state, at which point it would remain until some exogenous event would dislodge it and renew the cycle. The science depended on theory, moreover, because it lacked the empirical basis that had underlain traditional knowledge. Without theory it had a handful of anecdotes amid a scene that was changing far faster than it was.

But in its assumption of stability and predictable change, the Clementsian concept was fatally flawed, while for ecological communities that were mangled by settlement and subject to mass immigration by exotic species it was irrelevant. Lacking a firm baseline, however, made it difficult to know how to correct conditions, particularly when those past conditions were themselves the outcome of earlier disturbances. There was no naturally established restore point. Scientists could measure effects but could not say what the proper or natural state should be.

What drought did to the Great Plains, invasives did for the Great Basin, and both left Clementsian ecology in tatters. The Great Basin became the Dust Bowl of exotics. Instead of returning to a prior equilibrium, the Basin filled with newcomers that outcompeted the natives. The theory behind an avowedly applied discipline was too flimsy to cope with the consequences. Theory was hard to dislodge, however, for without it scientists were little more mentally equipped or able to speak authoritatively than were ranchers. The two groups became the yin and yang to the evolving understanding of what was happening in the Basin.

Much as ranchers learned to incorporate cheatgrass into their annual forage cycle, so researchers struggled to cope with what invasives did to their conceptual understanding of the world. For some ranchers, the insistent weed was a godsend that filled the critical gap in the annual cycle of herding, the need for spring pasture. For others, it was a noxious weed and a fire hazard of gargantuan proportions. So it also proved with scientific ideas about cheat. For some thinkers cheatgrass was an alien that had no place in the natural order of the Great Basin and should be eradicated. It had to be fought, uprooted, replaced. For others it was part of an uncontrolled experiment co-conducted by nature and humanity in which biotic communities were always sorting themselves out. Settlement had made the Great Basin into a spotty terra nullius that restless, pushy species had discovered and now colonized. Cheatgrass was something to live with until the future imported newcomers to replace it as cheatgrass had tumble mustard.

But what was the intellectual analogue of early spring pasture? The new science was in turmoil, not only because it was the equivalent of a startup company, but because it disrupted the prevailing notions of how ecological change occurred. Researchers responded to the challenge according to their institutional contexts, their disciplinary training, the particular patch of range and summer pasture they studied, and the times they studied it. It was hard to generalize, though nearly everyone did. The competing perspectives meant that appealing to "the science" did not clarify the cheatgrass crisis in the Great Basin any more than it resolved the prairie crisis in the Great Plains. It only codified into scientific language and management journals the confusions that were fast unfolding on the ground.

The Clementsian model was not the only one available to ecologists, but it was easily taught and it was attractive for management because it laid down baselines by which to measure change and because it prescribed what management should do to reestablish the natural order. Latent within it was a policy that said that noxious weeds like cheatgrass had to go, or at least be replaced by something tame that could fill the voided niche. It argued for eradicating the weed, downsizing grazing, and stopping fires. In reality, the best analogy for what was happening in the Great Basin was less a recurring succession, akin to the stages of settlement, each creating the condition for the next phase, each repeating the cycle of the last, than of a mining stampede that left behind the ecological equivalent of slag heaps, abandoned mines, and deserted towns. Restoration involved more than resetting the Clementsian clock. It meant rebuilding biotas.

Still, there was a need to understand, and the fast-morphing landscapes were the subject and stimulant for the emerging sciences concerned with the biota of the Great Basin. One emphasis was agronomic, aimed at increasing the productivity of commercial livestock, which meant attending to the range on which they fed. Another focused on wildlife, especially game species but later nongame as well (particularly those protected by the Endangered Species Act). Some of what preoccupied researchers were species like elk, a native, but which, thanks for disturbances, had exploded in population far beyond historic levels until it behaved like an invasive. Some were outright exotics like the chukar partridge, which found in cheatgrass an ideal forage. The chronicle was becoming as scrambled as the biota. Sagebrush expanded in some locales, and shrank against juniper and cheatgrass in others. What had been patchy burns in perennial grasses leaped into expansive cheatgrass conflagrations and stand-replacing burns in pinyon-juniper. What had begun as a regime upset by abusive grazing had metamorphosed into one unhinged by abusive burning.

There was no simple fix. Managing for any one species unsettled others. Applying science in the form of reductionism only tended to multiply unintended consequences. There was no intellectual model that could reconcile ecology, economics, politics, and culture when all, each on its own, as well as collectively, were not moving toward some stable end point but towards increasing disruption. Moreover, the record of science-based intervention does not make uplifting reading. Too many of the failures had the prevailing science of their day behind them. By the time, 20 years later, they would be shown to be wrong, the scene had deteriorated further. It was not just that the complexity was too great for an emergent science, which it was, but that the conditions were local and not easily generalized and that those conditions were changing; and in fact, they were changing due to treatments prescribed by science as well as by feral events and chance. Cheatgrass, particularly, altered its phenotypic expressions and probably its genotype. By the time the science got applied, the scene had changed. By the time the results were in, the science too had changed. Its continual adaptations made "the science" look less like secular revelation than a kind of traditional ecological knowledge for a digital age. Empirical experience accumulated over hundreds and thousands of years had been trampled away. Even the power of modern science could not replace it with usable know-how in a matter of years or decades.

Context mattered. Among scientists there were contests and confusions over scale, definitions, starting points. Lab results did not easily translate into the field; field studies in one site did not extrapolate elsewhere. Facts manifested themselves in particular settings in place and time: they were more than data yet far from transcendental truths. There are hotly argued opinions about what the historic fire regimes of the sage steppes were, whether sage was stable, whether pinyon-juniper was expanding.[6]

It would take more decades, and more consistent conditions, to sort it all out. But management needed to act now, and to act with respect to an invasive, disruptive species that not only quickly adapted to change but created those changes. So, likewise, modern science, having trailed settlement into the Basin, had naturalized, and was changing future conditions of understanding. It had replaced the former diversity of knowledge with a singular methodology. It was not going away, nor could it fill all the old niches. Small wonder confusion flourished about what was happening and how, or whether, to correct it.

The only points of consensus were that cheatgrass was expanding its range and that its fires had mutated into a menace.

FREE-RANGING FIRE

Fire protection came both early and late. Institutional interest arrived with the U.S. Forest Service, which oversaw reserves on many of the Great Basin's mountains. These held little commercial timber, but they did boast important watersheds and summer pastures, both vital to range livestock. On them the agency imposed its national fire policy: it sought to prevent fires from starting and to suppress those that did occur. (It had some success, which swelled the area of shrublands, which caused population surges among mule deer and elk, all of which became vulnerable to subsequent cheatgrass fires.)

With the 1934 Taylor Grazing Act, the Grazing Service—and after 1946, the Bureau of Land Management—had responsibility over the basins and steppes where cheatgrass was most widely distributed. Fire was not just something the agency had to deal with, one arrow in a quiver of problems. It conveyed a sense of crisis, which made it a brilliant fulcrum by which to build up an organization; it even came with its own emergency financing. The modern BLM fire organization emerged after the 1964 Elko fires; by 1969 the agency was willing to challenge the Forest Service as a putative equal, at least in the region, for which the Boise (later,

National) Interagency Fire Center was an apt symbol. The placement of a national facility on the fringe of the Great Basin was not accidental: the Great Basin (and Alaska) were the BLM's twin champions.

The push of expanding cheatgrass fires pulled the BLM fire program along with it. It's hard to imagine the BLM emerging as the fire behemoth it became without the Great Basin aflame. With the Alaska National Interest Lands Conservation Act (1980), it lost sole ownership of Alaskan fire. Its other holdings throughout the West, while large in aggregate, were checkerboarded and scattered. The core of its empire was the Great Basin, which helpfully burned. By the 1990s it was burning wider and more savagely. By 2015 concerns over the sage grouse, broadened the issue from the BLM to the national fire establishment. Between them invasives and the WUI helped revive suppression as a national strategy.

The short history of wildland fire policy and programs is that the biota was unhinged first by removing good fire, then by unwittingly promoting bad fire. Along the way, the Great Basin went from a minor feature in the country's pyrogeography to a black hole (or at least a blackened one) that was sucking in attention and resources from everywhere else, not so much a counterweight to the resource-draining WUI but a wildland twin. There was less geographic or bureaucratic room to maneuver, whether in response to wildfires or for remedial treatments.

That applies to ideas as well. Proposed treatments came with competing philosophies of environmental management that argued for and against direct remediation. Behind them stood a philosophical debate—an ethical one, in the end—about whether active management, informed or not by science, was appropriate.

FROM PLEISTOCENE TO PYROCENE

In 1932 G. D. Pickford of the Great Basin Experiment Station published what might be considered the foundational scientific paper on grazing, fire, and cheatgrass, but the popular perception came from Aldo Leopold who penned an essay, "Cheat Takes Over," destined for *A Sand County Almanac*. As Leopold saw it the situation was clear enough on the ground. What needed discussion was what to do about it.

Cheat had already become the cipher for ruinous invasives everywhere. The basics were all there, including the perverse way cheatgrass compensated for abusive grazing by providing some spring fodder and by reducing erosion. But mostly there was the compounding threat from fire. "I listened carefully for clues whether the West has accepted cheat as a necessary evil, to be lived with until kingdom come, or whether it regards cheat as a challenge to rectify its past errors in land-use. I found the hopeless attitude almost universal." Had he written 50 years later Leopold would have found a vigorous effort to rectify that past, and an almost universal attitude of hopelessness about the prospects.[7]

This time scientists were pitted not just against ranchers and hunters but environmentalists. The quarrel among the first could be argued quantitatively, over dollars and data. The quarrel among the second was about philosophy and ethics. It was about how to operate in a world that nearly a century of scientific research and active management had not made better, and often made worse. It was about what kind of world we wished to inhabit and who we wished to be as inhabitants. Was it better to accept the invasives and the Great Basin they created? or to find a less obnoxious exotic? or to incorporate the newcomers, however offensive, into the pattern of future life? Or was it better to work toward restoring the indigenous species, whatever the cost and whatever the collateral damages might be? To pull back from decades of well-intentioned but ineffective measures; to suppress fires; to let nature heal itself, as best it could? Cheatgrass would not self-destruct, it would not self-deport. Could native perennials, even without grazing, replace it? Or could it only be replaced by another invasive, one hopefully less obstreperous, but one perhaps more noxious?

These were not disputes likely to be settled by more data or appeals to the technical literature. But they speak, as research papers and policy tracts cannot, to the core of what the Great Basin would become. Whatever the resolutions, or muddling throughs, or the emergency reactions, it was likely that fire would be central to that world. The Pleistocene was yielding to a Pyrocene, the former ice ages to a fire age generally known as the Anthropocene. Whether combustion in all its manifestations would shape the landscapes as completely as ice and pluvials had earlier remains to be seen.

A WORTHY ADVERSARY

"I ADMIRE CHEATGRASS."

That's not a common sentiment. It sounds even stranger coming from someone who has spent a career grappling with what he calls the "invader that won the West." But if you are a competitor at heart, if you value tenacity, adaptability, and patience, if you accept that the nimble will prevail over the sluggish, then cheatgrass is not merely an ecological survivor or a stray gallows-humor success story but an extraordinary rival. It came. It saw. It conquered. And it did what few conquerors do, it stayed. We judge heroes by their villains, we value contests by how hard fought they are. For a land manager in the Great Basin, cheatgrass can seem an ultimate challenge. A weed that can remake whole biotas, that can reshape the largest federal land bureaus and bend research organizations to its will, can also shape a life. So it happened with Mike Pellant.[1]

He was born and raised in central Kansas, far from the seat of cheat. Fifty years earlier cheatgrass had pioneered sites from Reno to the Wasatch Front. Twenty years before he arrived in Concordia in 1952, it had exploded over the northern half of the Great Basin, and G. D. Pickford was publishing the first serious scientific studies of grazing, fire, and cheatgrass at the U.S. Forest Service's Great Basin Experiment Station in

Utah. By World War II locals accepted cheatgrass as rural Africans did malaria and sleeping sickness. It was an indelible part of the countryside. It could no more be eradicated than mosquitoes.

In 1949 Aldo Leopold's posthumously published *A Sand County Almanac* included an essay that had made cheatgrass the model and type for vile invasives, the ultimate symbol of what could go wrong by the silent aggression of an overlooked plant. With what, in historical perspective, seemed breathtaking speed, it had leaped from the straw of thatched roofs in Europe to whole landscapes of the West. Worse, "it is impossible fully to protect cheat country from fire," and the shrinking native landscapes, now assaulted by fires, were "the key to wildlife survival in the whole region." Cheat was here to stay. It couldn't be plowed under, grazed out, poisoned, or burned out, all of which only fed its spread. So the infested regions had found ways to make "the invader useful." It couldn't be beaten. It wouldn't self-extinguish. Leopold discovered among residents a profound fatalism. Cheat was awful, but it was shrugged away as better than grazed out pastures and bare ground.

Yet the postwar era witnessed largescale experiments in land use and science applied to the service of political economy. The BLM, formed by administrative fiat in 1946, sought to rehabilitate the western range that was its bureaucratic homeland. That meant taking on the weed that had taken over so much of the landscape. It was the postwar era that saw the last great dams raised, that chained off pinyon-juniper woodlands, and that converted large swaths of sagebrush steppe to crested wheatgrass. That last scheme served a double purpose. It transformed sage-infested wastelands into productive range and it stymied the growth of cheat. Crested wheatgrass made good forage, and if properly grazed, it could replace—serve as an ecological surrogate for—the native perennials lost to the era of bad grazing. The problem was that the project was expensive and it destroyed native vegetation that a not-too-future generation would regard as intrinsically valuable, a future more inclined to value native birds than imported livestock. The wholesale conversions rose and fell while Mike passed through elementary school and high school. By the time he entered college, an environmental movement was finding value in the once-despised sage steppe, much to criticize in the commodity-driven management of the public domain, and suspicion of a science-commercial complex that had loosed toxins, clear-cut wildlands, and sprayed DDT everywhere. The efforts to extinguish cheatgrass looked like the misguided campaign to eradicate fire ants.

Mike's most enduring education came on his grandparents' farm, 50 miles from Salina, the nearest town. He spent four summers there, from the age of 12 to 16. His mother had come from a family of 10, and he joked that with all the siblings gone, his grandparents needed another source of "slave labor." But they were delightful summers. He learned to observe nature closely. He grew to love plants. He thrilled to the annual spectacle of useful greenery reclaiming plowed earth. His grandfather had a self-informed but serious land ethic, and Mike absorbed it. Meanwhile from his father, a junior high school science teacher, he acquired a curiosity about nature, an appreciation for science, and a predilection to problem-solve nature's questions by scientific principles.

Then, for his final two years of high school, he spent his summers as a biological technician for the Army Corps of Engineers at the Wilson Reservoir, a subtle segue from a naturalist perspective to a scientific one. When he went to Fort Hays State University he majored in botany and continued to work in the mixed-grass prairie at Wilson Reservoir; and there he became involved with prescribed fire. He pursued a master's in botany at Fort Hays, with an emphasis on range science, and used that rich prairie and its burns for his thesis. He even experienced his first escape fire. By then he was ready to leave Kansas for a wider world.

He got more than he perhaps bargained for. His advisor had spoken of the BLM, which suited Mike's passion for range management. In 1976, with his MS thesis in final review, he was appointed range conservationist at Monticello, Utah. It was, he recalls, a moment of "culture shock." The Colorado Plateau, with its incised gorges and mountain laccoliths, was unlike anything he had seen, and small-town Mormon country unlike the rural communities he had known growing up. He wasn't in Kansas anymore.

In retrospect it turned out to be the "best thing" imaginable for his career. He had to deal with badly disturbed landscapes. For two years he had to inventory the damages and inspect up close—a countryside so unlike rolling prairie that it might have been on Mars. He learned to listen to and understand people whose background differed from his own. Over the next two years his survey transitioned from degraded rangelands to landscapes deranged by uranium mining. Call it extreme restoration.

Then it was time to move on and move up. He applied for a supervisory range job in Boise; the interview took place in the foothills, amid lands converted to cheatgrass. Here Mike experienced his second culture shock. It was, he thought, the "ugliest country" he had ever seen. But as cheat had taken over the sage steppe, so a new aesthetic gradually overtook him, and he came to appreciate what he saw. He naturalized to the place. Boise became home. And he came to respect cheat as a competitor.

It was fire country of course, much of it burning within sight of the National Interagency Fire Center (NIFC), officially inaugurated 11 years earlier. Inevitably he became involved with the BLM fire program, first as a resource advisor, eventually in suppression. It was what you did. And fire, he quickly learned, paid much of the freight. The overtime and hazard pay from fire season was a welcome, and expected, annuity—not quite on the scale of Wall Street bonuses, but a Boise echo. Whatever its liabilities, cheatgrass boosted spring forage for ranchers, and so, it seemed, it did also for the summer wages for firefighters.

Two views emerged. People concerned with land health worried about the cheat infestation and its knock-on effects. People hired for fire crews saw it as endless employment, a problem but not one that aggressive suppression couldn't hold. There were many fires, but few conflagrations. Then the fires got bigger and meaner. Mike recalls a conversation with an old rancher at Mountain Home, gleefully explaining how his father had "made this land with a match," letting cheat replace sage, before some years later admitting that the overall deterioration and loss of forage had reached a point that was driving him out of business. So, too, fire managers came to recognize that the fire scene had mutated into something monstrous. Wildfires began burning into the outskirts of Boise itself.

The new regime demanded more serious rehabilitation. Here Mike found his calling. Coaxing good green out of black and bare soil was what he loved. He became a technical expert in restoring grass and shrubs. But emergency response was only first aid. What the land needed was an ecosystem wellness program.

In 1985 he became coordinator for an Intermountain-wide experiment in planted fuelbreaks and postfire rehabilitation that went under the name greenstripping. The teen farmer from Kansas now oversaw row crops of crested wheatgrass and forage kochia intended to break the sweep of cheatgrass-powered fires in the Great Basin. The width of the strips varied by location from 30 to 400 feet, but Idaho's averaged 300 feet, often adjacent to roadways. By 1987 the program had created a handbook that identified greenstripping criteria and protocols and with the help of Idaho's congressional delegation got funding for pilot projects. Four years later results looked sufficiently promising to scale the program up into Utah, Oregon, and Nevada.[2]

Was it a brilliant structural solution to cheatgrass, or an act of growing desperation, a final roll of the dice in a game of gambler's ruin? Fuelbreaks work best when they are built into the design of a planting program, not when they are retrofitted onto deranged landscapes. The green forage drew livestock and wildlife to them, which, if unchecked, reduced their value and invited roadway accidents. The strips had to be extensive enough to prevent cheatgrass from doing an end run: they were only as strong as their weakest link. And like all fuelbreaks, greenstrips had to be maintained and would survive against the odds only by persistent funding. Congress, and the public, love to build roads; they are reluctant to maintain them. The same considerations applied to postfire rehab. Sowing crested wheatgrass could replace cheatgrass, but at the cost of substituting one monoculture for another, and one that had, again, to bow to a grazing regimen that was unpopular with ranchers but because it supported herding was unwelcome to many environmentalists.

This was no final solution, but at best a chock block behind the wheel of a semi that could keep the truck from rolling further down the hill. Greenstripping was a stabilization measure. Rehabilitation was not restoration. Cheatgrass remained on the land, and in its genomic versatility was expanding into saltbrush. It had a persistence and patience and nimbleness that could dance rings around Congress and bureaucracies and disregarded the ambitions of ranchers and environmentalists alike. It milled with other noxious weeds. It defied simple control by grazing. It had an even more implacable ally in wildfire. Like a software virus it had infected and rewritten the operating system for Great Basin ecosystems. It took on all challengers.

By the mid-1990s, after ecosystem management had become the mantra of the federal land agencies, Mike Pellant summarized the options for coping with cheatgrass. They were the stuff that only a tenacious optimist and competitor could read without sinking into a salt flat of despondency.

The first need was to reduce cheat's spread. This meant keeping grazing from destroying the perennial native grasses where cheatgrass was present but not (yet) dominant. It meant interrupting the positive-feedback loop that made fire and cheatgrass so lethal. Where cheatgrass was already dominant, it meant replacing it with a more manageable surrogate, which is to say controlling it before planting by deep disking, spritely timed burning, seasonally targeted grazing, and herbiciding, all of which inflicted their own collateral damages. Then it meant reseeding. What had become the tradition plant of choice, crested wheatgrass, could compete if planted in cheatgrass-cleared sites and could support cattle, but it did nothing to help the indigenous vegetation and was tarred with the memory of mass conversions in the 1950s and 1960s. All this was a stop-the-bleeding first aid. The ultimate goal had to be restoring something like the presettlement biota, which among other impediments demanded sufficient native seed, which didn't exist. The solutions, Pellant noted, "are few."[3]

What had changed since the alarms of the 1930s was that cheatgrass had claimed yet more lands (and more kinds of lands) and that scientific interest and management experimentation had expanded. The first symposium dedicated to cheatgrass had been held in Vale, Oregon, in 1965. Some 89 participants heard 20 presentations. When the community next gathered, in Boise in 1992, 340 participants heard 140 presentations. Since then symposia have come on a cycle not unlike cheat fires.

So, too, the cheatgrass problem on the land had magnified by an order of magnitude. Cheatgrass had gone from an unfortunate exotic that had become vital as spring range to an existential threat to the entire ecosystem. What had changed was the escalating character of wildfire. The summer the Boise symposium met, the Foothills fire had blasted over 257,000 acres around Mountain Home, the largest wildfire in Idaho since 1910, its plumes tauntingly visible from NIFC. Those fires made urgent what had been a remote and for most of the country a merely symbolic nuisance and ecological parable.[4]

The fires grew. In 1999 they exploded over nearly 1.8 million acres, the Great Basin's answer to the Big Blowup. The 1986 Dorsey Butte fire that rampaged over 17,000 acres of the Snake River Birds of Prey National Conservation Area had stunned observers. The complex of burns that massed into the 1999 season increased the scale by three orders of magnitude. The Dun Glenn Complex ran to 361,000 acres; the Sadler Complex to 209,500; the Corridor 171,442; the Battle Mountain 156,958. The Mule Butte fire blackened 138,915 acres; the Slumbering Hills 103,641; the Jungo 83,939; the Denio 77,244. Even the smallest of the big fires clocked in at 50,000 acres.[5]

Twenty years after the Boise symposium (and half a dozen successor symposia) cheatgrass threatened to unhinge not only the biota of the Great Basin but the BLM and through it the national fire establishment. Cheatgrass had become a biotic equivalent to the WUI: it was sucking in resources that should have gone to general land management.

Yet after 1999 there were those who refused to walk away from the challenge. There were programs for postfire rehab, for weed control, and, after the 2000 National Fire Plan, for fuels treatment. But emergency fire rehabilitation only becomes active after fires. Weed funds only apply after the weeds have become serious. Fuels management has no metric for invasive grasses. Suppression could only hit the ground when flames did. All are fundamentally reactive. None address the comprehensive and, by national standards, alien character of the Great Basin. What was needed was a holistic strategy that could eradicate, rehabilitate, and restore entire ecosystems.

What resulted from the 1999 conflagrations were two reports, *Out of the Ashes* and *Healing the Land*, that served as blueprints for the BLM to create a Great Basin Restoration Initiative (GBRI). The existing approaches each had their own mission and monies. The GBRI had to integrate the old approaches and find new ones, all under a novel conception. The vision of an appropriate response moved from greenstrips to whole landscapes. Mike Pellant moved with it. He had served on the post-1999 season review committees, upgraded his standing as emergency stabilization and rehabilitation coordinator for Idaho to the national office, proselytized on behalf of the GBRI, instructed at the BLM National Training Center, and ultimately became coordinator for the GBRI.

That required the skills of a ringmaster: there were programs to catalyze the right science, programs to build up native seed banks, programs for hands-on rehab and restoration, programs to design suitable equipment, and the tireless massaging of institutional arrangements. The USGS, the U.S. Forest Service, the Natural Resource Conservation Service, the Western Association of Fish and Wildlife Agencies, Joint Fire Science Program, regional universities, and of course the BLM all had commitments in research and practice. It was the kind of task Mike relished. In a way his entire career had escalated along with the cheatgrass-fire problem. He kept pace, however much the country lagged.

By now events were far outstripping the capacity to respond. Invasive grasses had leaped, as an issue, far beyond the borders of the Great Basin and were unmooring even the strategies that had emerged from the fire revolution. The 653,100-acre Murphy Complex (one of three adjacent complexes that totaled three million acres) kept the Great Basin in the national cauldron that was the 2007 season. Three years later, on the centennial of the Big Blowup, the Fish and Wildlife Service determined that the cheat-fire dynamo was "a significant threat to the greater sage-grouse" and considered whether "protections under the Endangered Species Act are warranted." That promised a political firestorm if the sage grouse did, in fact, get listed.

The GBRI suddenly found itself close to the center of the national fire establishment. The only hope of saving the sage grouse (and sparing agencies from a listing) was to save what habitat remained intact and roll back what was lost. Out of what had become an ecological equivalent of plowed ground it was necessary to grow new and improved ecosystems. How to do this administratively, politically, and technically, amid the still-festering injuries of settlement and the fast-marching insults of the Anthropocene, was a challenge only an inveterate competitor—a glass-half-full kind of guy—could savor.

On January 5, 2015, Secretary of the Interior Sally Jewell issued Secretarial Order 3336, *Rangeland Fire Prevention, Management and Restoration*. The Rangeland Fire Task Force charged with responding returned its final report in May. *An Integrated Rangeland Fire Management Strategy* incorporated a century of fire management, 20 years of understanding of what ecosystem management meant, and a decade of experience in collaborative enterprises. It was an "all hands, all lands," risk-based, science-informed, landscape-scale program to check, and if possible roll back, cheatgrass.

It was not a task for people who like marching orders that tell them what to do and how to do it and who could imagine analogues of ticker tape parades to recognize their success. There would be no final victory. Cheatgrass would remain; fires would continue; there would be no armistice, no end point. But with luck and pluck the devastation could be arrested and some of it nursed back to health. That had been Mike Pellant's career. Soon it would be someone else's job.

In December 2013 Mike retired. It didn't stick. His job was too much who he was, and the job was not done. He returned for a three-year appointment, relishing the unique opportunities to coordinate so wide a program and to mentor future staff, which he preferred to do on the ground. When that tour ends, he hopes to continue consulting on select projects that most interest him, push for experiments in targeted grazing, and train future BLMers.

He leaves with some parting thoughts.

He had seen fires since his days at Wilson Reservoir. But those of recent years are to the old burns what buffel grass is to Bermuda. After the Murphy Complex, he spent half a day flying the burn in a helicopter, speeding over black as far as the eye could see. It was an epiphanic moment, the shock that we are entering "unchartered territory." The uncertainties are greater than we thought, the future more portentous than we conceive. As much as a cause of change, the monster fires are a symptom of it. They have become the "ultimate challenge" for the Great Basin.

They can be fought of course, but at the cost of a huge buildup of suppression resources, and with no prospect of final victory. They simply kindle too easily, move too fast, escalate too abruptly, and every burn spurs on the next. The emerging treatment of choice across the public lands of the West—managed wildfire—will not work in cheatgrass since any kind of fire can cycle back in a positive feedback. The only hope of controlling cheat-powered fires is to control the land, which is say, the cheatgrass itself. Fire and cheat have come to resemble an M. C. Escher drawing in which one stairway rises to another, and another, and around until the series ends paradoxically at the place of origin.

Still, there are points of light amid the gloom. Targeted grazing, Mike believes, must be part of the solution; in a sense, prescribed slow combustion might do here what prescribed fast combustion does elsewhere. There are prospects for biocontrol like the fungus known colorfully as the "black fingers of death." There are cases of extensive cheatgrass die-off, perhaps from soil microbes. The causes are not understood, but those patches, some large, are points for intervention, a kind of emergency rehabilitation not unlike replanting after a wildfire. They can add up.

Yet it isn't enough to dampen or even extinguish cheatgrass. The crisis developed because the native perennials were destroyed by bad grazing. The ecological niche still exists. Something will fill it. Where a flower blooms a weed cannot, and where the flower fades, weeds will rush in. The replacement might resemble the indigenous bunchgrasses and co-sustain sagebrush, or it might resemble cheatgrass and wipe out the sage. That leads Mike to his nightmare scenario.

Cheatgrass is not the toughest, nastiest, most loathsome weed lurking in the ecological shadows of the Great Basin. It has some value as forage, it stabilizes soil, it's adaptable; until the big fires broke out, we managed to live with it, if unhappily. The horror is that the day might come when some future Aldo Leopold visits the Great Basin and finds the cheat suppressed, a more demonic invasive thriving, and the scene so far gone and the despair among residents so profound that they look back longingly to the era of cheatgrass, not perhaps as a golden age, but as one at least of golden hillsides that shone brighter than the biotic lead that succeeded it.

Not all newcomers are toxic. Some, Mike Pellant among them, naturalize successfully, and make their habitat the better for their presence. But all carry with them an evolutionary past that shapes their disposition to plant or plunder. William James noted that a "man's vision" is the great thing about him. A part of what makes the long- and multiserving coordinator of the nation's efforts to cope with cheat is the vision cultivated by a youth on his granddad's farm eagerly awaiting, with a patience born of hope, the new greenery that would arise from bare ground.

MUSHROOM CLOUDS

On May 23, 2016, preparing for President Barack Obama's visit to Hiroshima, the *New York Times* ran a feature on the celebrated photograph of the city's destruction under a towering mushroom cloud. The wrinkle, however, was that the plume did not come from the atomic bomb but from the innumerable fires kindled by it that merged into a colossal mass fire. The diagnostic mushroom cap was in reality a pyrocumulus. Kevin Roark, a spokesman at the Los Alamos National Lab, which made the bomb, noted that the photograph's plume was in fact larger than any mushroom cloud from even the most powerful nuke ever detonated, a "thousand times stronger" than the one exploded over Hiroshima. For 70 years the photo had been miscaptioned.[1]

That will not come as much of a surprise to the wildland fire research community, because for almost that long fire has been recognized and studied as a fundamental consequence of thermonuclear weapons. Blowup fires came to be described in terms of so many Hiroshima-type bombs and identified by their mushroom-cloud tops. The fires ignited by nukes are distinctive only by their source. These critical experiments, and those behind the aerial incendiaries that preceded Fat Man and Little Boy, were tested in the Great Basin. White Sands, New Mexico, was sufficiently empty to test one bomb. When it decided to transfer tests from Pacific Isles to mainland America, the Defense Department needed still greater emptiness. It found it in Nye County, Nevada.

Probably humans have weaponized fire as long as they have held a torch. From the *Iliad* to the Battle of the Wilderness in the American Civil War, fire was common. Even when not set deliberately, once gunpowder made firearms abundant, fires were frequent on battlefields. Too many ignition sources, too much tinder on natural landscapes. The fog of war was most often smoke.

It's an old narrative, but one the 20th century polarized. World War I saw plenty of firepower, but little open burning along the Western Front because artillery had pulverized the land into mud. World War II, a war of movement, released fire. Incendiaries abounded, often delivered by air, culminating in the atomic bombs over Hiroshima and Nagasaki. The destruction the nukes wrought, however, was less than "conventional" fire weapons dropped onto Kobi, Tokyo, Hamburg, and Dresden. Blast and burn were a lethal combination. The Strategic Bombing Survey after the war reckoned that fires wrought many times more casualties than blasts. The next war, they concluded, would likely be a fire war.

The weaponization of science is nearly as old (think Archimedes). Science demanded experimentation. During the war, the War Department had built mock cities to explore how to burn them better but the fires were potentially so vast they sprawled over landscapes. Horatio Bond, chief fire engineer for the National Fire Protection Association, a tutor to the Strategic Bombing Survey, had written in 1943 that "a very definite policy should be developed for looking into the use of fire as a weapon." After the war, he documented the destruction possible in a classic study, *Fire and the Air War*. This was a vital topic for research. The only agency that routinely experienced big, propagating burns on such a scale, however, was the U.S. Forest Service.[2]

Fire research had early been part of the agency's mission. In 1911 William Greeley had asserted that it could be studied much like silviculture, and for decades it was a colorful but tiny branch of applied forestry. The needle shifted with the appearance of Wallace Fons and George Byram, a mechanical engineer and a physicist, respectively. The push came after the war when a planned reorganization of fire research in 1947 coincided with military interests. It was enlisted in the Cold War against the "red menace." What had existed as a niche subculture of forestry now mingled shoulder to shoulder with high-octane physicists, chemists, meteorologists, and mathematical modelers. It was a heady time, particularly after the Soviet Union exploded a bomb and the Korean War began. "We have been given a classified military project of broad scope," A. A. Brown, then head of Forest Service fire research, exulted. The research "enables us to do a lot of highly technical work we need for our own programs" and "has already gained a great deal of prestige for the Forest Service and is regarded as highly successful by the Military." The Office of Civil Defense added funding as well. The National Academy of Sciences National Research Council established a Committee on Fire Research. Following the Cuban missile crisis, more money flooded in, and American fire researchers joined with Britons, Canadians, and Australians for field trials and modeling.[3]

The Forest Service had plenty of reasons to pursue fire science. It just had never had much money to invest. The Department of Defense (DOD) and Office of Civil Defense monies seem a tipping point both in institutional scale and disciplinary themes. It opened three regional fire labs, at Macon, Georgia, Missoula, Montana, and Riverside, California, and then hired scientists well outside traditional forestry—men like Richard Rothermel and Hal Anderson—to staff them. The Riverside Lab oversaw a large series of field experiments, Project Flambeau, to study the properties of mass fire. Part of the payback came during the Vietnam War when several mass fires were attempted to deny cover to the Viet Cong. One worked; most simply towered into pyrocumuli, and then rained on themselves. Forest Service involvement smelled like scandal. The funds dried up. Then the Reagan administration oversaw a massive transfer of research dollars from civilian to military purposes and labs. If DOD wanted fire research, it would use military labs or contractors. By the mid-1980s wildland fire research was on the ropes.[4]

The era of mutual assistance probably did more for wildland fire research than it did for the military. The future would not feature fire wars, but it did hold more fires than ever. For military strategists fire was too erratic, too little controllable. For wildland agencies, however, the research promised to bring better control to a phenomenon that came with or without declarations of war.

Blast and radiation were predictable. Both followed from a single detonation; both obeyed known laws of physics. But free-burning fire was not similarly predictable. It integrated its surroundings. A bomb at one time could set whole landscapes aflame; the same bomb at another time could fizzle. A bomb imposes; a fire synthesizes. Though fire is the most damaging because it can propagate, it is also the less controllable for the same reason, which is why the military abandoned it. Blast and radiation are over in seconds. A landscape fire can burn for months.[5]

The strands of this particular narrative braided together nicely in 1961, a year after the Missoula Fire Lab opened and a year before the Cuban missile crisis rekindled interest in thermonuclear mass fires. In a year that seemed big with fires at the time but that now seems quaint, three stood out. The Sleeping Child fire in Montana was a classic backcountry burn. The Harlow fire through the foothills of the Sierra Nevada roared through a largely rural landscape and community. Both were traditional kinds of fires that burned in expected ways.

The stunner, though, was the Bel Air-Brentwood conflagration that burned through ritzy mountain exurbs of Los Angeles. It announced a new kind of fire, one that seemed to advertise the kind of lethal destruction forecast by military strategists. Twenty-five years later it would acquire the geeky name for which it would be the type specimen: wildland-urban interface fire. In retrospect, Bel Air-Brentwood stands as WUI One.

It also illustrated the way fire would differ from other nuke effects. The inexpungable image is of Willard Libby, Nobel laureate in physics, best known for his discovery of radiocarbon dating, but also chief scientist and shill for the Atomic Energy Commission, where he became notorious as an advocate of fallout shelters. Nuclear fallout was ephemeral and predictable, he insisted, which meant that nuclear wars were survivable with proper protection. He even built a "poor man's shelter" out of bagged dirt and railroad ties for $30 in his backyard. A few weeks before Bel Air-Brentwood he was photographed peering out of the shelter in white tuxedo and black bow tie. But while fallout might be predictable, fires were less so. The conflagration burned Libby's house and incinerated his shelter.[6]

The panic over the Cuban missile crisis led to a big bounce in research funding through the Office of Civil Defense. Even so, field trials decoupled fire from nuclear blast. In 1963 the threat of fallout was enough to advance the Partial Nuclear Test Ban Treaty, which moved future tests underground. The Forest Service proceeded with large-fire field experiments, but the dissociation was becoming intellectual as well. The question of starting mass fires became less useful than the question of defending against them. The military concluded that blast and radiation could be weaponized, but not fire. A bomb would blow up whatever it landed on. A fire might go out, or it might spread. It might even turn on its igniters, even if they wore a tux and clutched a Nobel Prize.

Detonating nuclear warheads over Frenchman and Yucca Flats was a novelty in the Great Basin. Fires were not. The flash burns from the rising fireballs of bomb tests were not the fires, or the fire regime, the southern Great Basin was accustomed to or expected, but in some respects fire is fire, and the biota accommodated.

The Nevada Test Site (NTS) encompasses two desert regions and their vegetation types. The south is Mojave Desert, a dry expanse of blackbrush and winter annuals. The north is Great Basin Desert, higher, slightly wetter, rich in sagebrush and pinyon-juniper woodlands. In presettlement times it's unlikely that either burned routinely, only when winter rains brought a rich supplemental crop of annual grasses and forbs to help propagate fire, or after enough centuries had passed to allow the brush to grow close and the woodlands to thicken to the point they could carry flame with a stiff wind. That changed when settlement, primarily by livestock, crushed the old regime and allowed for *Bromus* to infect the site. During the 1940s and 1950s some areas were burned to promote better forage than afforded by the indigenous shrubs, which were indigestible by cattle.[7]

Before aboveground tests began, four species of brome laid claim to the site. *Bromus rubens* (red brome) dominated in the south; *Bromus tectorum* (cheatgrass) to the north. They thrived on disturbed sites. In 1962 the NTS arranged with Dr. Janice Beatley for a series of permanent plots by which to measure vegetation change; the next year the Partial Atomic Test Ban Treaty forced explosions underground. The active fuels were red brome and blackbrush, which burned more readily than sagebrush. An inquiry published in 1966 noted that both bromes were well established. Both promoted fire. Both appeared to be "stable," having achieved most of their expansion during the era of their introduction (ca. 1920s), well before the NTS was gazetted in 1951. Red brome was expected not to change much "in the future" by the disturbances (including fire) accompanying test programs at the test site. Cheatgrass was set to explode, however, as "much virgin vegetation" was "decimated or totally destroyed in the construction of new road systems and test installations."[8]

At first blush it's an unsettling thought. Could it really be that establishing the Nevada Test Site did less lasting ecological damage than overgrazing before and new construction afterwards? Is regional ecology following the pattern of WWII urban conflagrations in that "conventional" incendiaries wrought more damage than nuclear ones? Twenty-five years after the original monitoring plots, the Department of Energy arranged for a series of studies at NTS to satisfy National Environmental Policy Act requirements, which in turn led to an ongoing ecological monitoring and compliance program, with special emphasis on threatened and endangered species. Heavy rains in the spring of 2004, which led to a flush of grass, inspired an extensive survey of wildland fire hazards, which continued through 2006.[9]

The conclusions based on 40 years of data will sound familiar to anyone knowledgeable about fire in the Great Basin. Three invasive grasses were redesigning fire regimes: Arabian schismus at lower elevations, mostly in blackbrush; red brome at lower to middle elevations; and cheatgrass at middle to higher elevations. Lightning and vehicle ignitions each accounted for roughly 40 percent of ignitions, with 15 percent resulting from "military and security training," and another 4 percent from unknown sources. In all the NTS averaged 11 fires a year, each burning an average of 239 acres, with most of the burned area from a handful of fires that flash through red brome and cheatgrass after heavy winter-spring rains. The fires were "costly to control and to mitigate." Most of the ecological damage occurred in blackbrush, which yielded to other, more fire-hardy species, and there were economic losses in infrastructure ("powerlines and communication structures"), not to mention the price tag for suppression.[10]

Like many Great Basin newcomers, the nukes had blasted the land and then left. The enduring crisis was ecological, caused by pyrophytic grasses, climate change, and the construction habits of an industrial society. The lasting blowup burn would not come from mushroom fireballs but from the mushrooming combustion of fossil fuels.

DEEP FIRE

For millions,
for hundreds of millions of years
there were fires. Fire after fire.
—GARY SNYDER, "WILDFIRE NEWS"

IN THE YEAR 02000 CE a fire—believed to have been kindled by lightning just before midnight on July 25—started on the south side of Mount Lincoln in the Snake Range.[1] Steep terrain and tricky winds, plus extensive fires elsewhere in Nevada, argued for monitoring rather than suppressing the fire, which reached an acre in size the next day. It grew slowly for several days, then absorbed larger chunks. By August 3 it had spread to 1,250 acres. By the 23rd it had dropped off the western cliffs of Mount Lincoln and entered the valley below. There it met a cache of dry combustibles, and simmered before boiling up the slope of Mount Washington, billowing skyward in a vast plume that dwarfed the peaks. It overran the St. Lawrence Mine, purchased the year before by the Long Now Foundation as a potential site for a 10,000-year clock. It burst over the summit where it flash-burned a grove of ancient bristlecone pine. Seven incident management teams wrestled in sequence with the burn, the last departing on September 7, with the fire 90 percent contained.[2]

The clock and bristlecone move the Phillips Ranch fire from the realm of anecdote to apologue. It's impossible to look on those scorched sites and not meditate on fire and deep history.[3]

The Snake Range in eastern Nevada is a good place to look. It contains four of the five tallest peaks in Nevada. But it is not the height of the peaks that matters: it's their depth in time. That's true for its biota as well as its rock, and it's true for its peoples. It's a place Stewart Brand has called "timeless," but only because most humans think in terms of hours and weeks, and their digital machines in nanoseconds. We're a species for whom a lifespan of a century is an occasion for wonder. Even a millennium seems beyond our temporal reckoning.[4]

The larger narrative over the last 10 millennia is slippery because there are so many moving parts and they often move with extraordinary power and speed. The climate wavered and warmed, then cooled slightly. Glaciers shriveled to token cirques. Immense lakes—the Great Basin equivalent to subcontinental ice sheets, some 28 million acres in all—drained or evaporated. Half of ancient Lake Bonneville, originally the size of Lake Superior, mostly emptied over Red Rock Pass in the space of days. The flora scrambled across the new landscapes in the biotic equivalent of the mining rushes of the 19th century. Some 35 genera of mammals (mostly large: think mammoths) disappeared. The biota not only moved around and across basins but up and about the ranges that flank them.[5]

No less remarkable, through nearly all these wild upheavals, there were people present. The human saga is a story of movement because the Great Basin is a place not notably kind to fixed habitations and settled peoples. The earliest human narrative may date back 14,000 years. Clovis people arrived a thousand years later. Others—how many is unknown—followed. The Fremont peoples probably migrated in from the Colorado Plateau as early as 00400 CE before they decamped wholesale by 01400. By then the ancestors of the present-day Shoshone, Goshute, and Paiute had reached the Great Basin.

Emblems of their appearance had preceded them, but Europeans arrived tangibly in the late 18th century CE. The pace quickened as Americans filtered into, then swarmed over, the Basin. The mountains then witnessed a historic cavalcade of peoples, moving in, passing through, departing, and more recently preserving. The markers of time have shortened, not only because more records exist but because the tempo of change has quickened, no longer keyed to the ponderous rhythms of climate but to the teleconnections of distant economies, beliefs, and institutions.

Most of the Snake Range was gazetted into the public domain in 01891 CE as a forest reserve, later reorganized into the Humboldt National Forest. The lower reaches of Mount Wheeler contain an elaborate subterranean cave system, Lehman Caves, named after Absalom Lehman, the rancher who discovered them and became their promoter, opening paying tours in 01885 CE. The caves became a national monument in 01923 under the jurisdiction of the Forest Service. Ten years later responsibility was handed over to the National Park Service. In 01986 the monument and some additional lands, 77,000 acres in all, became Great Basin National Park.[6]

Throughout, there had to be fire—nature would see to that. Fires must have burned amid the forests that lapped between glaciers and pluvials, and then across grass, shrub, and woodlands that claimed lands vacated by ice and lake. Surely, there must have been some basal rhythms to burning, though the biota might still be quivering from all the massive shocks of the Pleistocene. Like the land under ice and water, which continues to flex upward in rebound once that burden was lifted, so the biota may yet be responding to the abrupt changes in that past climate, even as it must react to the sudden changes now underway. How many of the fires recorded over the past century are in some way legacy burns? How many are early-adopter harbingers?

But people also carried fire and put it on the land both purposefully and as an inevitable spoor of their presence; anthropogenic fire may be humanity's first camp follower. Migrating peoples added hunting, warring, trapping, mining, farming, and herding, and they introduced species, both deliberately and accidentally, including disease-spawning microbes. They rearranged the places and tempos of burning. The old rhythms must have found themselves syncopated, collated, isolated, quickened, dampened, and generally unhinged if not scrambled. A detailed study of fire scars and stand ages mumbled that "a distinct within-fire-shed contrast in fire frequency was difficult to explain without invoking the possibility of spatially-variable human-caused ignitions." In other words, people interacted, with fire, to reshape the natural scene. The study noted that a major inflection point—what for fire might be the equivalent of the onset of a glacial epoch—occurred in the 19th century CE.[7]

The fires that reached the ledgers of the 20th century CE are not likely those that characterized the presettlement era, though just when the first tendrils of European "settlement" arrived is a fraught topic. Trade goods, steel knives, iron pots, horses, and communicable illnesses like cholera, measles, and smallpox could precede and remake societies, and so remake landscapes, decades or even centuries before white Europeans officially recorded first contact. "Presettlement" makes an unstable baseline, related by unreliable narrators.

The fire history of the Snake Range is typical of that for the Great Basin's sky islands. An analysis of fire-scarred trees carries the story back to 01538 CE, two years before Francisco Coronado's expedition through the Southwest and 229 years before the Escalante-Dominguez expedition blazed the Old Spanish Trail down the eastern Great Basin. Most scars, however, track to the mid-19th century CE. The majority occur in limber and ponderosa pine, the rest in Douglas fir, white fir, and pinyon pine. Low-severity fires burned on average every 11 years in ponderosa (2- to 22-year range), and 19 years in mixed-conifer (1–62 years). In pinyon-juniper, fire return intervals vary by the degree of grassy understory, but a conservative estimate is that fire occurred every 15–20 years.[8]

We know the most recent fires from formal records kept by the Forest Service and National Park Service. A large fire (200–300 acres) broke out at the time the Forest Service assumed control, then a few arson fires blackened 150–200 acres, then almost nothing. Lightning became the only ignition source, kindling an average of three small fires a year. The Park Service acquired control just as the West generally began to swing into drought and an era of larger fires. Most of its notable fires, however, began on private or bordering lands and burned into the park such as the Phillips Ranch fire of 02000 and the 164-acre Border fire in 02006. In August 02016 lightning started the Strawberry fire that blew up into 4,700 acres on the north foothills of Mount Wheeler and became notorious when a falling tree killed a Lolo hotshot. But over a century of detailed records roughly 5 percent of the park has burned—not much even for a mountain range in the Great Basin.[9]

Fires take their character from their contexts. The generally complex assemblages of forests on the Snake Range make for equally complex patterns of burning. The Snake Range is a sky island large enough to accommodate many refugia. Likely, it's a palimpsest of natural history, and if so, that applies to its fire history as well. How far today's regimes deviate from presettlement conditions is unclear, and perhaps unknowable. Fire's regimes must have been as varied and no more settled than the riotous movement of flora, fauna, and people. They remain unsettled today.

Bristlecones are the oldest known trees on Earth. The youngest old bristlecones match the age of the oldest sequoias. Not all bristlecones are ancient, but all of Earth's longest-lived conifers are bristlecones. The oldest, named Prometheus, dated 4,862 years—2,100 years older than Hesiod's eighth century BCE account of the mythic Prometheus in the *Theogony*.[10]

The oldest bristlecone groves in the Snake Range—one on Mount Wheeler, and a larger one on Mount Washington—date to almost 5,000 years; the oldest tree, to 5,600 years. They have flourished within two frames. One is climate—the warm altithermal that ended about 6,500 years ago, and the more recent warming of the past century. The other is geology. In the Great Basin plants grow in a broad zone between basin playa and range treeline, or between the salt left by Pleistocene pluvials and the ice left by glaciers. Bristlecones grow near the upper limit, near the cold. Mostly they grow in mixed association with other trees. They please the mind with their age; they please the eye with their sculptural forms, like flames petrified into wood. Their fire ecology is barely understood. The fire history of the Snake Range is not much better known.

The patriarchal bristlecones survive because they live in an ecological sanctuary so hostile to life that the rest of the biota including predatory beetles and blister rust, competitors and predators, and rival trees have been filtered out. They have little ground litter, and grow well apart, so there is scant occasion for fire. Bristlecones will thrive nicely where limber pine and Engelmann spruce do, but bristlecones in those habitats are unlikely to live to old age, which for Great Basin bristlecones is about a thousand years, much less achieve patriarchal status. They find sites where nothing else can survive. But also critical to their longevity, they seem not to senesce. They don't biologically age. Instead, the ancients die from cumulative physical insults, which for many ultimately means that erosion exposes and kills their roots. If their habitat is framed by salt playa and stone summit, so their life span is also set geologically, not biologically.

Fires burn subalpine forests, and bristlecone with them. The trees can survive light-severity surface fires, but not much more, and so resemble a high elevation version of pinyon pine. The contrast with sequoias is striking. Old sequoias thrive in robust habitats; old bristlecones, in harsh, unforgiving ones. Redwoods live long because they accommodate fire. Bristlecones live long because they avoid it. Since the legacy groves burn rarely, their fire history is little known. But they do burn.

The Phillips Ranch fire stunned observers because it burned a large swathe through a grove of bristlecones. Such burns couldn't be common, or else the grove would not be old. Part of the explanation may, again, be the association with other species. For such fires not to have happened, fires could not have erupted up Mount Washington as they did in July–August 02000 CE for several thousand years. The bristlecone didn't change: its surroundings did. More savage fires were possible, most likely because regimes around it had been perturbed over the past century. A century is not long for a patriarchal bristlecone. But it's plenty of time to alter fire regimes downslope.

Equally astonishing, bristlecone appears to have regenerated preferentially in the burned area. How? Why? It's not known, but speculation suggests that once-occupied sites became available, and that the Clark's nutcracker was critical in distributing seeds. Perhaps, too, there was some genetic memory still coded from the times when bristlecone and limber pine dominated the flora of the Basin and had to accommodate fire regimes very different from those on Mount Washington and the quartzite summit of Mount Wheeler. Until ancient groves burned, research wasn't possible. Now, however macabrely, it is.

Still, the oldest bristlecones span half the length of the Holocene. They seemingly defy ecological time. They can't defy their geologic boundaries.

Fire raging forest or jungle,
giant lizards dashing away
big necks from the sea
looking out at the land in surprise—
fire after fire. Lightning strikes by the thousands, just like today.

The proposed Long Now clock would double the age of the oldest bristlecone. Ten thousand years back spans the Holocene, ten thousand ahead would presumably absorb the Anthropocene, or if the human hand shaping today's Earth proves too ephemeral, whatever geologic epoch will subsume it as the deep rhythms of the Snake Range would an afternoon's thunderstorm.

In 01999 CE the Long Now Foundation purchased 180 acres, including the St. Lawrence Mine, on the flank of Mount Washington with the intention of constructing a clock that would run for 10,000 years. It would serve as a tangible symbol to encourage thinking about the long term. Like ancient bristlecones it would survive because it would occupy a sheltered site, shielded from natural and human disturbances (fire among them). Within a year fire blew over the mountain.[11]

Ten thousand years is nothing for fire, which has thrived on Earth for longer than 42,000 times the length of the Holocene. Fire predates the genus *Homo* in all its avatars. It predates cheatgrass, sagebrush, and the Eocene origin of bristlecones. It predates the genus *Pinus*. It predates grasses and forbs. It predates beetles. It predates the evolutionary history of everything presently living on the Snake Range. It has outlived five glacial ages and five mass extinctions. It has outlived the creation of the range itself. It predates almost all of the rocks that make up the range.

It burned through the Pleistocene, while glaciers came and went, and talus, lacustrine gravels, and alluvium slid down the slopes. It burned through the late Tertiary conglomerates that eroded out and redeposited old Paleozoic rocks. It burned while lakes laid new deposits, and feldspar-rich and biotite-rich volcanics, rhyolites, and granites formed. It burned while the granites of the Cretaceous and Jurassic congealed. The Snake Range lacks any Triassic or Permian rocks, but fires scored trunks in Arizona's Petrified Forest that date from that era. The Permian was a time rich in oxygen, vegetation, and fires, probably the most burned era in Earth time. When the Ely limestone was laid down, earthly fires moved into uplands. When the Chainman, Joana, and Pilot shales were forming, fires burned in mires and the tropics. When the Simonson and Sevy dolomites began, fires first appeared, as plants, then forests, formed.

Yet, like bristlecone pine, fire is ultimately bounded by the deep-time evolution of life, which is to say, by geology. When the rocks that layer the lower Snake Range were quartzite, limestone, and shale, there were no fires. When the Osceola Argillite, Shingle Creek shale, and McCoy Group quartzites formed, there was only a thickening of oxygen in the atmosphere, a product of life in the oceans, the only life that existed. The geologic record goes back further, beyond even that border, but Earth at this stage in its history is indistinguishable from other dead worlds. Without life there cannot be oxygen, and without life on land, there is no fuel.

Fires—so many fires. Fires in glacial eras, fires in high oxygen eras, fires in eras of sparse fuels. Fires in ferns, fires in bryophytes, and fires in the first angiosperms. Fires when diplodocus browsed. Fires when ancestral mammals appeared. Fires in swamps that became the coal beds of the Pennsylvanian. Fires in prelapsarian savannas. Fires that burned as soon as life colonized land, and have never stopped. Fires that are reburning the fossil charcoal of Pennsylvanian coal. Fires that will burn until terrestrial life vanishes, until planetary history erodes away its roots, like the dragons and serpents gnawing at the three roots of the Yggdrasil. So many fires. So many fire regimes. So many possible futures. So many choices for the Earth's keystone species for fire to make.

What is the proper fire regime for the Snake Range?

With its life-zone stratification and terrain-sculpted nooks, the Snake Range holds landscapes and niches, all with distinctive regimes, that stack vertically in short order what would otherwise span across hundreds or thousands of miles of geography. But that shortening of space is also a shortening of time. The mountain replicates the regimes of centuries, millennia, even epochs. The current regimen, which seems so out of whack, is a distortion of less than a century. Today's blowups are whispers compared to the fires that no doubt blew over what the Pleistocene's ice sheets and pluvial lakes didn't submerge. Today's most expansive megafire is barely a tick in the pyrochronometer of the mountains' deep time.

But we don't live in deep time—can barely comprehend its scope. We live in a present that spans a handful of years, and can anticipate a future of perhaps a few decades. Amid the forecast upheavals of the Anthropocene, today's fires are a puff of snowflakes from a coming blizzard. Even the Long Now's admirable appeal to long-term thinking projects across 10 millennia, probably only 5 percent of humanity's past, and 0.004 percent of the Pleistocene. Compared to the antiquity of fire, that number becomes vanishingly small, an infinitesimal of pyrohistory.

The fires of today are the fires we must manage. Most fire managers have working horizons of three years—this year, last year, next year. They must cope with the fires they face, not those of the Miocene or of a predicted post-Anthropocene epoch. "Think global, act local" also has its historical analogue: Think long term, act now. Fire rarely allows us the luxury of simply contemplating.

The myth of Prometheus tells how fire came to people, and so to Earth. Curiously, perhaps, there is not one myth but many. Hesiod tells one version. Plato tells one. Others tweaked the basics to account for some particular feature or other, or more correctly to illustrate some ethical or political point. They are less retold myths than reshaped parables. The story—or what by now may be the myth—of the Prometheus bristlecone pine also has many variants, but they tell how we should relate to that Earth, which inevitably will speak to how we should handle our unique firepower.

The basic elements are these: In the summer of 01964 Donald R. Currey, a graduate student from the University of North Carolina, wanted to use tree rings to estimate the age of glacial retreat on Mount Wheeler. He took cores with a Swedish increment borer. Bristlecones were understood to be old—this based primarily on studies from the White Mountains in western Nevada. But he stuck a borer—there may have been two—in the tree he labeled WPN-114, but which locals had earlier named Prometheus. He asked for permission to cut the tree, which would allow him to recover the borer, better date the rings in the complex tangle of trunks, and complete the science. The request went up the administrative channels of the Forest Service, then the responsible agency. The authorities agreed. A crew trekked up with saws, and when one member refused to cut, returned the next day, August 7, and cut several slabs. When the rings were finally counted, it was discovered that Prometheus had lived 4,872 years, and a recount suggested a number closer to 5,000, maybe 5,200. Prometheus was by far the then-oldest known (once) living thing. Protest over what many deemed an inexcusable act of vandalism boiled over.

For meaning, chronologies (which is what tree rings are) have to become narratives, which is to say, they have to acquire a theme to organize events over a designated span of time. No narrative is inevitable: each narrator can tell the story according to his own perspective and purposes. If the narrative is history, the narrator can't make up anything and can't leave out something whose omission would alter the outcome, but otherwise is free to shape structure and meaning. If fiction, parts can be tweaked, added and omitted, for thematic clarity or aesthetic effect. If myth, fable, or allegory, the elements can assume a more abstract quality, in which nominal facts become symbols. What matters is that the resulting text is true to its genre and resonates with its readers. What complicates matters is that often even the supposed "facts" can be suspect. Facts, too, take their character from their context.

The Prometheus story has, like its referential myth, many versions. Michael Cohen has identified five "predominant" ones, which like root sprouts compete to become the primary trunk. Currey's version appeared in *Ecology*. His justification for cutting resided not only with the stuck borer (which could not be replaced easily), but for better access to the data, which the complicated growth habit of the tree made cumbersome, along with the need to finish his project. He was, after all, doing science. He concludes that "possibly no other living species presents such accessible long-term evidence relating to its biogeographic history and to the environmental histories of its sites," not only showing no remorse but appearing to invite others to do the same. Data is where you find it. Darwin Lambert, an advocate for transferring jurisdiction to the NPS, saw the Prometheus bristlecone as a martyr. Keith Trexler, an onsite witness and chief naturalist at the monument, interpreted the incident as reflecting the different values of two agencies. Galen Rowell, writing for the *Sierra Club Bulletin*, imagined a collusion between a commodity-blinded Forest Service and an instrumental science—the incident occurred, revealingly, the same year Congress passed the Wilderness Act. Charles Hitch, president emeritus of the University of California, also saw a clash of values, but argued for the free inquiry of scientists, even when errors can from time to time occur.[12]

The recounting of the events around the felling—the chronology, if you will—is secondary to the investment of meaning given them. In that sense the saga of the Prometheus tree remains a cautionary tale. We can hardly understand those events without their accompanying story; and this is no less true for fire history. How, then, can we tell the story of fire through deep time and know its place in our time? What is the purpose of fire history at all?

The value of placing fires in deep time is that it reminds us that fires change with their circumstances, that our understanding of fire also changes with circumstances, that only a few of those circumstances are under human control and people are unlikely to agree on them, that today's solutions can become tomorrow's problems, that there is no final resolution to fire. Fire will go on, with or without us.

Like the story of the Prometheus bristlecone, the saga of fire in the Snake Range, or for that matter fire on Earth, will have many narratives. More data will alter those stories, but data is just data, no matter how big or digital. Facts don't speak for themselves, are not even understood as facts without some setting. Meaning requires a frame, whether as narrative, aesthetic sensibility, or ethical conviction. But even for natural phenomena, even for something on the scale of climate, those frames are not set simply by nature. Through their firepower, people are moving them. This is not to say that those processes are under human control, just that they are susceptible to human disruption. Worse, the same agents destabilizing them are the ones telling the stories to explain what is happening and what it means. At this point understanding starts to resemble a Möbius strip.

Our formal understanding of wildland fire is pitifully small. The first American scientific paper on fire ecology was published in 01910. The first Forest Service lab opened in 01959, five years before the Prometheus bristlecone was felled in the name of science, and 41 years before the Phillips Ranch fire flashed into the Mount Washington grove. Moreover, what modern science has studied are much disturbed landscapes over short intervals of time. Most investigations last the one or two years allotted to graduate students. In truth, humans know a lot about fire—it's part of our species heritage, we've lived successfully with it for 200 millennia. But most of that traditional knowledge was lost in the chaos of American settlement and the various suppressions characteristic of fossil fuel–powered industrialization. What we have today are fragments, like the shards unearthed at a Fremont pit house. We hardly know the fire history of the Snake Range, and that for only a century. We don't really know much. We won't catch up. We won't get ahead. And the world we study is changing rapidly. We'll never know enough.

What caused the Phillips Ranch fire to burn into bristlecone? Was it just a chance event, a 5,000:1 bet that beat the house odds? Or was it the result of land use changes, including recovered woods after mining and fire exclusion over the past century? Or did it happen because, thanks to humanity's combustion of fossil fuels, the climate inflected in ways that added punch to the fire, like an atlatl leveraging a spear? Those geologic borders that once hard framed the Prometheus bristlecone have become malleable under the fussy hands of humanity; and in the end, people even felled the tree, and then offered explanations for what they had done. Which narrative we choose matters because we will base future decisions on that understanding, and that choice will enable or shut down other choices to come. It's easier to break than to build. It's difficult to create what we wish for but simple to disrupt the present or derail the future.

Those most passionate about global change, and the Four Horsemen of the Anthropocene, argue for a no-analogue future in which the past provides no prescriptions. Even the Long Now clock will record the time to come, not the time passed. In truth, a knowledge of fire history won't tell us what to do now. Its deepest purpose is different. Technology can enable but not advise; science can advise but not choose. The value in an appreciation of deep time is that it can counsel us how to make better choices by means of narratives that speak to virtues such as prudence, humility, compassion, courage, and grace under pressure, character attributes that might well be described as timeless. An appreciation for deep fire history won't tell us with mathematical rigor how to make landscapes more resilient. It might help prevent us from cutting down the future in order to retrieve today's increment borers.

I have to slow down my mind,
slow down my mind
Rome was built in a day.

OUTLIER

STRIP TRIP

IT COULD STAND as a dictionary definition of hiding in plain sight. At 7,900 square miles the Arizona Strip is nearly as large as Connecticut and Delaware combined. It contains nine wilderness areas, three national monuments, a national forest, an Indian reservation (Kaibab-Paiute), parts of two national recreation areas, and a patchwork of public and private lands. It abuts two of the most celebrated tourist sites in the country—the Grand Canyon and Las Vegas. Yet it might as well be on a Jovian moon.

Geologically, it contains the last of the High Plateaus running south from Utah. There are four of them, stepping down from east to west. The highest, most famous, most accessible, and least typical of the Strip is the shallow dome of the Kaibab, whose east and south flanks were sculpted into the Grand Canyon. To its west—what is understood as the real Strip—lie the broad, somnolent Kanab and the narrow, violent Uinkaret plateaus. The most western and forlorn is the Shivwits, wide and featureless, whose canyon gorge is a series of terraces, fit mostly for cattle, the occasional mine, and dead-end roads.

Here history and geography fuse. The Grand Canyon, passable only by a footbridge through the inner gorge, diverted or dammed settlement. The southward flow of colonizing from Mormon Deseret spilled to the more passable corridors to its sides, crossing the Colorado River to the east at Lee's Ferry near the Arizona-Utah border or streaming down the Old

Spanish Trail along the Wasatch Front to Pearce's Ferry on the west. The Nevada-Utah corner is further isolated by the Virgin River gorge. The Strip became a slow historical eddy filled with transhumant cattle (or later on the Kaibab, transhumant tourists) and a few postfrontier homesteaders.

What made a marginal landscape into the Arizona Strip were the unintended consequences of a political process that had severed the land from Utah and Nevada and given it to Arizona, which in turn found it split between two counties. Two-thirds of the Strip belong to Mohave County, with its seat in Kingman, and the remainder within Coconino County, seated in Flagstaff. Grand Canyon National Park and the Kaibab National Forest, each of which have large holdings north of the river, have their headquarters south of the Strip. The BLM administers its lands from St. George, Utah. The Strip is a political no-man's-land, the Empty Quarter of America's Empty Quarter. For a long time it was best known for the polygamous communities that persisted in remote indifference from any reach of law. When a group of brownshirts in cowboy hats staged a Sagebrush putsch at Malheur National Wildlife Refuge for 40 days in 2016, their leaders, Ammon Bundy and LaVoy Finicum, hailed from the Strip. The Strip is so isolated that it became a site for reintroducing California condors.

For the American fire community the Strip distills the lesson that significance is more than a matter of size or celebrity. Much as the remoteness of the Mogollon Mountains invited the first primitive area, the model for wilderness, so the isolation of the Strip prompted a national experiment in landscape-scale restoration. For a while the core woodlands of the Arizona Strip, the Area 51 of fire management, became the epicenter of a national debate about rehabilitating fire-famished forests. For a decade it was as much a ground zero for field testing the tools of wildland fire management as the Nevada Test Site was for nukes.

That happened because the Strip was a cameo of fire history in the American West, though one with historical and ecological quirks of its own, and one that came with layers of isolation that, paradoxically, could entice a program of active experimentation.

The lower shelves, broad and elevated above the norm, were grassy and shrubby, grading in higher reaches into pinyon and juniper woodlands. The critical site was the Trumbull Range, a chain of volcanic cones, piles, and peaks that flanked the eastern faultline of the Uinkaret Plateau. In Paiute *Uinkaret* means "place of pines." These were ponderosas that grew in classic southwest fashion—huge, clumpy, a score or two per acre amid grasses. Even before the pine arrived, roughly in sync with the modern climate 6,000–8,000 years ago, people lived here. Likely they moved seasonally as water and favorable habitats became available. They dug irrigation canals and farmed Toroweap Valley before abandoning the enterprise in the 13th century. Lightning, torch, grass, brush, and forest coevolved.

By the time American settlers arrived in the mid-19th century, Trumbull's fire history, as recorded in scarred boles, spoke of routine fire every 4.5 years and widespread fires every 9.5 years. These are minimum values, but they correspond nicely with chronicles throughout the Greater Southwest. Then that pyric saga ended. The last outbreak occurred in 1863.

Colonizing Americans had arrived, along with sheep and cattle, axes and sawmills. While they often followed similar seasonal rhythms, they introduced new elements and had novel outcomes. Instead of harvesting pinyon nuts, they harvested whole ponderosa pines, and instead of hunting deer and rabbits, they loosed livestock. Instead of adjusting settlement patterns to the cadences of climate, they doubled down to compensate. The result was an ecological shockwave. By 1870 landscape burning ceased to be a routine feature of the biota. Save for the occasional outbreak of a few dozen acres, fires disappeared. Instead of stoking flames, combustibles fed livestock and steam engines.

What followed happened all over the arid West, though it was told here with a local twang. The luscious range crashed, eaten out and trampled. By 1880 Captain Clarence Dutton, then detached for duty with the U.S. Geological Survey, viewed the panorama from Pipe Spring, ranching's point of entry. "Ten years ago the desert spaces outspreading to the southward were covered with abundant grasses, affording rich pasturage to horses and cattle. Today hardly a blade of grass is to be found within ten miles of the spring, unless upon the crags and mesas of the Vermilion Cliffs behind it. The horses and cattle have disappeared, and the bones of many of the latter are bleached upon the plains in front of it." The close cropping might have been enough by itself, "as has been the case very generally throughout Utah and Nevada," he argued, but a decade-long drought had also settled on the scene like a down blanket, smothering everything beneath it, and the double blows were too much for the native flora to survive.[1]

The assault paused, then continued. Cattle, horses, and sheep from Utah; then Texas longhorns; then tramp flocks; decade after decade, another grazer appeared, taking what water and forage had survived, like waves of prospectors sieving through the slag heaps of earlier booms. In May 1933 a party from Grand Canyon National Park traveled to Toroweap. They found that the landscape was "badly over-grazed, erosion is severe, and loco weed and other undesirable indicators are present." The next year the Taylor Grazing Act was passed, closing the public domain to further entry. An estimated 100,000 cattle were still on what remained of the range, along with an unknown number of sheep and feral horses. The Grazing Service created by the act brought some order to the pandemonium and reduced livestock numbers. In 1946 the Grazing Service was corralled with other Department of the Interior strays into a Bureau of Land Management. Today, 117 ranchers graze 15,000 cattle under permit.[2]

In 1879 a dairy operation located at Trumbull. But herders generally were moving their flocks to the higher forests during the summer, and there they met loggers. The Trumbull Range timber was good—a small mill was in place in 1870. The problem was, as always, isolation, the long haul to market. That objection vanished when Mormons began construction of the St. George Temple, found the Trumbull timber suitable, and built a Temple Trail to haul the wood by ox team. A steam sawmill speeded the process. When the Temple was dedicated in 1877, the Mormons sold out. The felling and milling went on. What was happening in the range was occurring in the woods as well. If the havoc was less severe, it was because the U.S. Forest Service imposed some institutional order after 1905.[3]

This was the third leg of the land-history stool, federal management, which added formal fire protection at least in principle. It meant that fires big enough to be seen would be fought, but more, perhaps, it meant that the ancient legacy of human burning would cease. The Strip's peculiar isolation, however, made its lands awkward to actively manage. Whatever the agency or administering unit, the Trumbull Range in particular remained an outlier, handed around an extended family of federal land bureaus like an unwanted orphan. The agencies seemed to change tenure as often as ranchers swapped spreads.

The Trumbull Range and Mount Dellenbaugh (an outlier of an outlier, located on the Shivwits) were incorporated into the Dixie Forest Reserve in 1904. The Wilson Administration dropped the Parashant (Dellenbaugh) division from the national forest system. In 1924 the Mount Trumbull holdings were transferred to the Kaibab National Forest. Ten years later Grand Canyon National Monument was proclaimed for the lower reaches of the range and the Toroweap Valley where it empties into the Grand Canyon proper. In 1974 the Kaibab transferred the Trumbull unit to the Bureau of Land Management. Mount Trumbull and Mount Logan wilderness areas were gazetted in 1985. In 2000 the expansive Parashant-Grand Canyon National Monument swept up the mountains into its domain, acting as a kind of local Taylor Grazing Act to support an environmental ethos appropriate for a service and amenities economy.

The response of the woods was not, like the early range, to shrivel into dust but to morph into a thickening scrub. The forest overgrew, in the manner typical of the region. Instead of open glades, brush and dog-hair thickets flourished, and instead of the classic colonnades of old-growth yellow pine, forests became the woody "jungles" that tracked the failing health of Southwest ponderosa. Average tree density went from 22 stems per acre in 1870 to 349 by 2000. The exception was the rumpled summit of Mount Trumbull itself, which had resisted the logging technology of the early days and still held significant numbers of old-growth yellow pine.[4]

Meanwhile sparse local population increasingly slipped off to the towns and ran cattle as a side business, though in full costume. At times the Strip resembled a reenactment event, recycled from the 19th century.

Fire as an integral part of the Strip was gone. Lightning still kindled snags, and a few fires reached a few tens of acres, but they appeared as someone might stumble upon a basalt arrowhead or the wall of an Ancestral Puebloan ruin. They were relics of a former era. The ancient yellow pine ceased to record any further scars. When two "severe" wildfires broke out in 1996, one was 20 acres and the other 150. For a long time the absence of routine fire had gone unnoticed—fire's removal had been policy, after all. The consequences of fire exclusion elsewhere in the West sparked concern, however, and then alarm, and finally inspired a famous experiment under the direction of Northern Arizona University professor Wally Covington on the Gus Pearson Research Natural Area outside Flagstaff, Arizona, that sought to explore what it would take to restore fire in the pines.

Quite a lot, it seemed, primarily by mass thinning of the small-diameter intruders, followed by prescribed burning, and then probably by seeding to leverage the native grasses. But before the Flagstaff model, as it came to be called, could leap from its tiny prototype plots on the Gus Pearson, it needed a major field trial. It needed experiments not only to sharpen the science but to sculpt the protocols for operations: it needed a full-spectrum enterprise in how science and management might coevolve the theory and practice of ecological restoration on a significant scale. In 1995 Secretary of the Interior Bruce Babbitt, who had grown up on a Flagstaff ranch, and Arizona's senators, particularly Jon Kyle, met with Covington, and all eyes turned to the Trumbull Range as an ideal locale. The catalytic 1994 fire season (which had savaged areas around Flagstaff) spiked the discussions with a sense of urgency. The upshot was the Mount Trumbull Ecosystem Restoration Project in 1996, sponsored by the BLM in coordination with the Ecological Restoration Institute at Northern Arizona University and the Arizona Game and Fish Department.

The Trumbull Project was big science for its place and time. It sponsored a full-spectrum ecological CAT scan and laid out systematic plots for field trials primarily in ponderosa pine, but also in pinyon-juniper and cheatgrass, all with monitoring and control sites for various combinations of treatments, all at a landscape scale. For a restoration point the project selected 1870, when the long rhythms of fire had abruptly ended. Comparative studies were initiated at Beaver Creek Biosphere Reserve, and Grand Canyon's North Rim, both in Arizona, and at the Sierra Tarahumara and the La Michilia Biosphere Reserve, Chichuahua and Durango, Mexico, respectively. The Arizona Game and Fish Department signed on as a collaborator and added faunal studies.[5]

An environmental assessment began immediately. The scientific inventorying, and the site preparations for treatments, occurred mostly between 1997 and 1998. By 1999 the first data were in and reports were issued. Second entry treatments were underway by 2001. These proceeded more slowly due to costs, changes in national administrations, and adjustments based on early results. From the onset the project's design had insisted on "adaptive management" in which the original prescriptions, and those shaped by the research, would be modified as field conditions warranted. Experiments were field trials; field trials were experiments; they both coevolved. On some topics surprises were technical: restored fires did not behave as forecast. On other topics the snags were cultural and political, conveying disagreements over means and ends. A final report was issued in 2006. But some research persists, and updates continue to appear.

Where had the rubber hit the road and gained traction, and where had the wheels just spun in place? The surprises were many, but the designers had expected to be surprised. *The number, size, and arrangement of "replacement trees" that would recreate 1870 conditions had to be reevaluated.* Critics wanted more trees. *Gambel oak thinnings were scrapped.* The copses were valuable wildlife habitat, and their contributions to fuels around old-growth pines could be compensated for in other ways. *Control plots were expanded to accommodate birds. Cuttings burned too hot*, but smashing the slash, which reduced fire intensity, threatened soil compaction. *Native grasses, shrubs, and forbs did not return after slashing and burning, which led to reseeding, which failed*, until the treated areas were dragged to cover the seeds. What had seemed transparently obvious often turned murky. The "challenges" were endless. Insufficient funding and staff. Reliance on third parties for cutting and commercial use. "Excessive mortality" of yellow pines and Gambel oak from treatments. Continued grazing. Lack of native seed sources. Labor-intensive and expensive NEPA analysis. Complexities about maintaining every habitat for wildlife while doing the treatments. And that old incubus: "Addressing concerns of a variety of constituents and avoiding litigation."[6]

But the most vexing was the failure of restored fire to do what it was expected to. Prescription windows were too narrow, the lopped-and-left slash could burn too intensely, the fires encouraged cheatgrass, and, despite raking around their bases, more yellow pines died than anticipated. Surely, the millennial drought of 1999–2000 aggravated conditions, but protecting the old growth was the practical and ethical justification for the program. The point of restoration was to make those biotas more resilient and accepting of fire, which in turn would catalyze other vital processes. More troubling, perhaps, was the fact that the landscape—and this was an experiment in landscape-scale treatments—was a shambles of the treated and the untreated, a crazy quilt of retouched, untouched, and untouchable fuels. Large as the project was, it was not large enough to produce the protective buffers needed. Still, the program adapted, prepared for a second entry, and planned for larger plots.[7]

To the prime movers the results were good enough and the need to expand dire. Wally Covington urged that the BLM extend the treatments into the Mount Trumbull Wilderness to save its old-growth ponderosas. Here, one great idea met another. Restoration confronted wilderness, and wilderness won. Restoration is flexible, adaptive, learning by doing. Wilderness is nonnegotiable. It isn't based on experimental science or empirical knowledge from practice. It celebrates the wild, which transcends museum relics like Trumbull's many-centuries-old yellow pines. Restoration is based on the belief that we can use our science to intervene constructively. Wilderness ideology considers our science another potential form of meddling. Besides, the Mount Trumbull Wilderness had not been unhinged by logging and still retained many of its native grasses. Let nature sort it out.

Instead, the lessons went into the Healthy Forests Restoration Act and other programs for hazardous fuel abatements, and eventually the Collaborative Forest Landscape Restoration Program. They did not, paradoxically, expand into nonwilderness areas of the Trumbull Range. In 2000 much of the Strip, including the Trumbull Range not in Grand Canyon National Park, was put into a Parashant-Grand Canyon National Monument, which introduced still other values and points of friction. A phase three treatment plan was approved in 2008 and hovers in the wings. A Uinkaret environmental assessment, scheduled for completion in July 2016, targets 128,000 acres, of which 30,000 would be treated mechanically and 18,000 by prescribed fire. But little funding is anticipated even if operational plans based on the environmental assessment pass muster. What made the restoration experiment possible at Trumbull, the Strip's isolation, is also what makes continued restoration implausible.

Mostly the Trumbull forests will fall under national fire policy and practices, with local adjustments based on that remarkable era of the restoration experiment. Most probably something like managed wildfire will replace prescribed treatments. Between July 23 and August 10, 2012, lightning kindled three fires on Mount Trumbull. They were handled as a complex under a strategy of "resource benefit" previously known as wildland fire use. It was a wet monsoon, but the deep duff kept the fires smoldering. For a while crews raked around the bases of yellow pines to protect them from root scorch, and a few patches of pine torched, but then the crews backed off. The fires burned into December. Mortality was within historic ranges. Landscape patchiness returned. Bark beetles came, but so did the three-toed woodpecker, long absent from the unburnt Strip.

What of the future? The Ecological Restoration Institute wants a 20-year review. It would be an excellent capstone to an extraordinary adventure in American fire and a confirmation of the belief that if restoration could happen anywhere, it should be able to happen here. Whether that science might inform a new era of practice and policy is less certain. The opportunity to apply such principles in big projects—even on the modest scale of Mount Trumbull—may have passed, or at least passed by the Strip. It may be the best lessons get absorbed into larger agendas. It may be that a bold concept came and went as others on the Strip have before it.

In 1880 with Kanab as a reference point, Captain Clarence Dutton, along with William Henry Holmes, the expedition's artist, made a series of treks to understand the geology of the High Plateaus and their complementary Grand Canyon. From the Aquarius Plateau he pondered the limitless horizon and the unimaginable scale of erosion that had stripped away the Mesozoic into the Great Rock Staircase. At Zion he meditated on the fabulous sculpturing of cliffs. Then they went to Trumbull and Toroweap. They ended on the Kaibab at a narrowing peninsula Dutton named Point Sublime.

Trumbull and Toroweap offered two geologic lessons, one on volcanism, one on fluvial erosion. The two converged at the Canyon's rim, where the Uinkaret fault that traced the Toroweap Valley met a cinder cone and basalt dikes that spilled into a Canyon gorge that was itself the final incision of serial erosion. The tableau marked an extraordinary conjunction. "Apart from merely scenic effects," Dutton asserted, "it would be hard to find anywhere in the world a spot presenting so much material for the contemplation of the geologist." From a perch on Vulcan's Throne, Holmes drew one of the greatest scenes of Canyon art, *The Grand Cañon at the Foot of the Toroweap—Looking East*, while Dutton contemplated how to make sense out of the unprecedented scenery before him.[8]

His key point was that all the facts he had collected needed to be assembled if they were to achieve meaning. The whole—the "grand *ensemble* [italics in original]"—was what mattered. So for his monograph as a whole, Toroweap's dramatic features would become one theme of many synthesized at Point Sublime, and it was there, in the form of a triptych, that Holmes created the greatest single work of Canyon art, *The Panorama from Point Sublime*. Dutton's science has dated; Holmes' art has endured. So likewise the Trumbull science may expire, while the idea that inspired it may survive.

Such may well be the fate of the Mount Trumbull Restoration project. Like Toroweap for Dutton, it marks the intersection of large forces—in this case, of ecological science, politics, and wildland fire. It inspired a great work of artifice, perhaps unrivalled for its era. And it may end up not as a centerpiece but as a vital contributor to an ensemble of experiments as Americans struggle to live with fire. Like the politics that ultimately created it, it was an exercise in the art of the possible.

It's worth recalling that great innovations take time. Scanning the scene at Toroweap overlook, Dutton noted that "the human mind itself is of small capacity and receives its impressions slowly, by labored processes of comparison. So, too, at the brink of the chasm, there comes at first a feeling of disappointment; it does not seem so grand as we expected. At length we strive to make comparisons." So exalted were the hopes for the Trumbull project that a passing retrospective might leave a similar deposit of disappointment. We need time, we need comparisons. It took 37 years to translate Dutton's text and Holmes's art into national park status for the Canyon, and 52 years for Toroweap to acquire standing as a national monument. It may take decades before the full ripple and rebound effects of the Trumbull experiment make themselves known.[9]

For now, the Trumbull project may be best understood as part of suite, not sufficient by itself, but not derived from anyplace else. It's the Strip Trip of American fire. It was, in its way, a pragmatic vision quest to a difficult place that is hard to reach and perhaps even harder to know.

FIGURE 1. Utah invasives. Cheatgrass and juniper claiming a once-sagebrush plain.

FIGURE 2. Kolob fire, Zion National Park. A type conversion to cheatgrass appears to be underway.

FIGURE 3. Mesa Verde National Park. Juniper ruins among Ancestral Pueblo ruins.

FIGURE 4. Mount Trumbull Wilderness fire (2012), captured by Google Earth.

FIGURE 5. High plateaus and high valleys, large landscapes and small towns, looking west toward the Sevier Plateau, southern Utah.

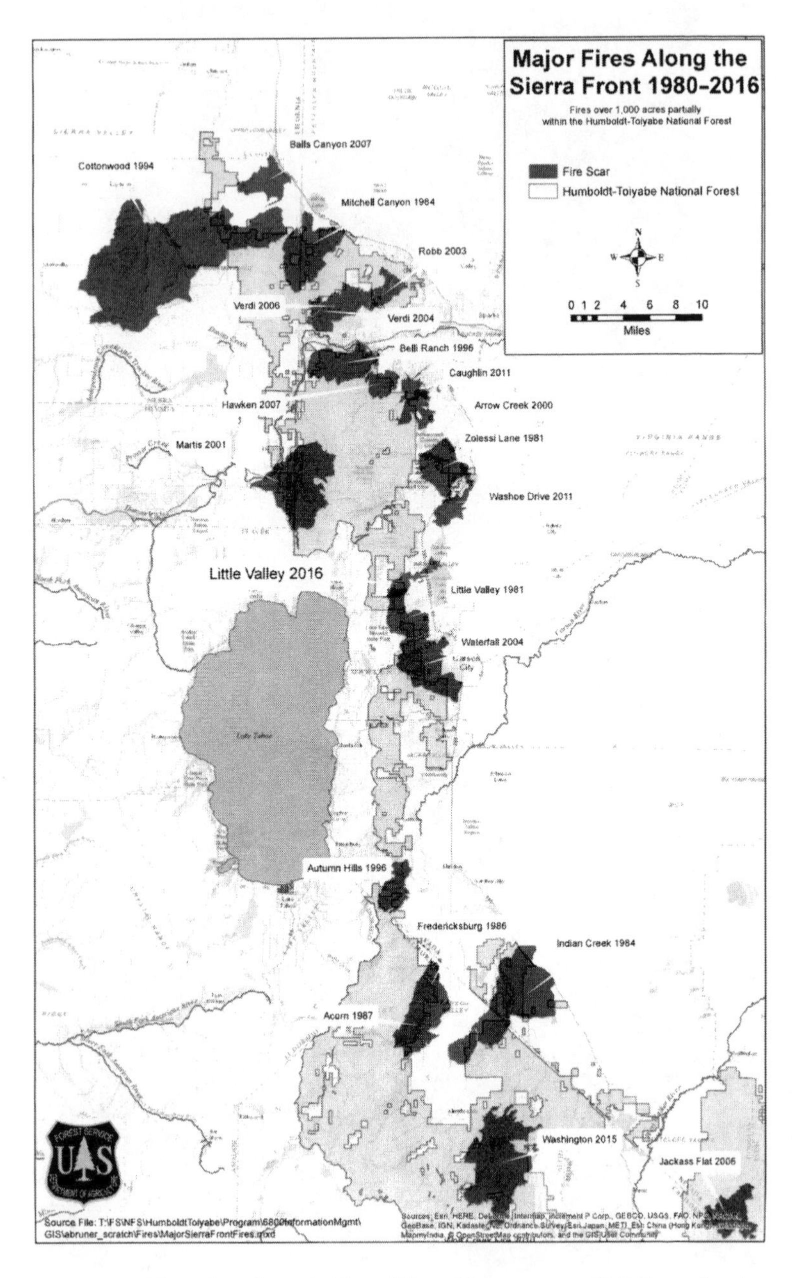

FIGURE 6. Major fires along the Sierra Front. Photo courtesy U.S. Forest Service.

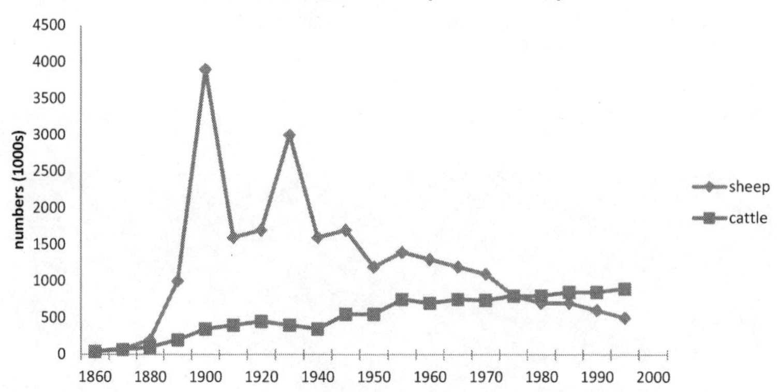

FIGURE 7A. Livestock in Utah. Data from Richter (2009).

Utah sawmill production (1890–2005)

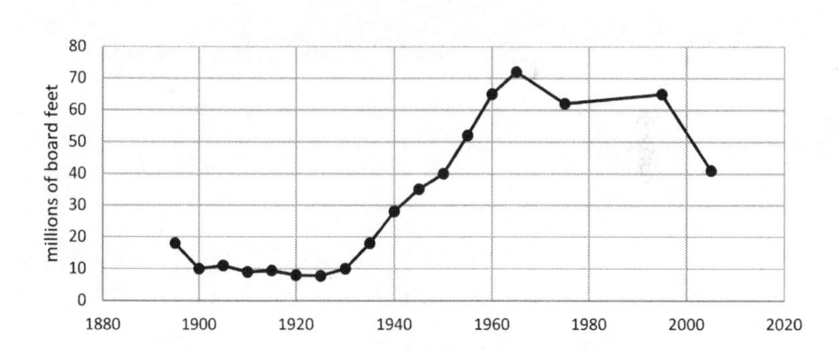

FIGURE 7B. Logging in Utah. Data from Richter (2009).

FIGURE 8. Pyrocumulus over Hiroshima, long incorrectly identified as the mushroom plume of the atomic explosion. Photo courtesy U.S. Army.

UTAH

ZION'S HEARTH

UTAH SITS CLOSE to the heart of the Interior West. It reflects both the region's divisions and its unities. Here its two grand physiographic provinces, Colorado Plateau and Great Basin, hump and hole, face each other across a fault line. Here America east and west joined, when in 1869, at Promontory Summit north of the Great Salt Lake, the Union Pacific and Central Pacific railways drove a ceremonial golden spike to their final rail. And here American religious enthusiasms met American secularism, each promoting a pattern of western settlement at odds with the other.

What made Utah different from Colorado and Nevada was the story of the Mormon exodus that brought a pioneering group of religious refugees to the Wasatch Front, and from there established a shadow empire, Deseret, throughout the West. These were settled communities, grounded in farming and close-packed villages, far removed from the riotous mining towns and far-droving stockmen that first populated most of the West, and then left it in slag heaps and weeds. This was the pilgrim as pioneer, the New England village, baptized into the City of Zion, gone West. The Mormon classic town was clustered, planned, "called" as a unit—the kind of close settlement championed outside the United States by Edward Wakefield and tried in New Zealand and South Australia, but that seemed to spark allergic reactions among westering Americans.[1]

For most of the Interior West the problem was to establish some basis for permanent settlement, an economy that did not rely on veins of ore that ran out suddenly or on sprawling herds that free-ranged on public land until a bad winter might kill them off or on bands of freebooting men more intent on a fast buck than communities with families, schools, and churches. That it solved this issue meant that Utah survived, and for some thoughtful observers it suggested a model that might be aptly applied elsewhere.

But what made that possible—settlement by a relatively homogeneous religious group—also made Utah a political problem. For all the traits of the America in which the Mormon prophet, Joseph Smith, had grown up and that the LDS church absorbed, the way those pieces assembled made them a "peculiar people." For all the emphasis on individual agency and self-reliance, on voting ("sustaining") decisions, and on replacing a permanent clergy with universal priesthood to all males over the age of 18, the political structure looked theocratic, the economy communitarian, and family life perverted by the practice of polygamy, which reformers paired with slavery as the "twin relics of barbarism." The problem in most of the Interior West was to rebuild out of the wreckage of first-wave encounters; the problem in Utah was to integrate the peculiar practices of its settlers into American society.

The Americanization of Utah was particularly tricky because Utah came with its own creation story. In this it resembled Texas, California, and Alaska, all state-nations. So, too, was Utah as the rump of Deseret until relentless pressure forced it to submit to national norms. It was founded during the Mexican War in what was a geopolitical no-man's-land until the Treaty of Guadalupe Hidalgo granted those lands to the United States and a people fleeing persecution found themselves roped back into submission. Separatist sentiment persisted, manifest in the Utah War of 1857 and most tragically in the Mountain Meadows Massacre. Utah Territory's claims across the Great Basin were steadily pared down. Its petitions for statehood—it had a sufficient population by 1850—were denied until LDS president Wilford Woodruff formally renounced polygamy in 1890. It joined the union in 1894. (By contrast Nevada was granted statehood quickly in 1864, but saw its population actually drop as the mines played out.)

Mormondom had a founding document in the Book of Mormon which tells a story of pre-Columbian settlement of the Americas, and whose translation by the prophet Joseph Smith led to a "restoration" of the Gospel. But the true creation narrative is the story of the trek to the shores of the Great Salt Lake. It's eerie how close that saga mimics the epic of Exodus. The persecutions by the state. The crossing of a providentially frozen Mississippi River. The long hegira in the wilderness led by a patriarchal Moses in the person of Brigham Young. The pronouncement by Young as he looked over the promised land between the Wasatch and the Great Salt Lake that this was indeed the place. So similar are features—which were recognized by the participants themselves—that Bernard DeVoto, an Ogden native, relied on metonymy and called the trekkers "Israel" in his retelling of the national movements that made 1846 a "year of decision." Outsiders were Gentiles. Even the Wasatch Front echoed the geography of the Holy Land, with its desert and mountains and its Dead Sea and Sea of Galilee and river Jordan to join them. No less than their Scripture, the Mormon story seemed a recapitulation, if not (certainly to believers) an outright restoration.

Utah literature has meant, directly or indirectly, Mormon literature, and Mormon literature has long fixated on the trek and its consequences, the literal gathering of Zion. Utah's major writers have been historians. Most Gentile commentators typically note such oddities of Mormon theology as the doctrine of eternal marriage or practices like the Word of Wisdom that proscribes tea, coffee, and alcohol. But Mormonism is a practical religion. Its testimony is the experience of the lived life; the Mormon ward (parish) is a village for raising children. Most literary efforts tend toward family histories and journals and of course retelling the saga of the trek. July 24, Pioneer Day, rivals the Fourth of July as a holiday.[2]

There is less interest in fiction, and when it happens, it focuses on the conflict between individual and group, and between group and nation. In Mormon literature the epitome of the former is Maurine Whipple's *The Giant Joshua* (1942), in which individual agency and belief conflict with group pressures and revealed precepts, in this case bolstered by the presence of polygamy. Gentiles might look for models to Mark Twain's observations in *Roughing It* (1872), which admired Mormon tenacity while dismissing its theology; Arthur Conan Doyle's *A Study in Scarlet* (1887), in which he introduced Sherlock Holmes with a backdrop in Salt Lake City; or Zane Grey's near-calumny *Riders of the Purple Sage* (1912). Mormons seemed an exotic society, easily adapted to a century that welcomed discoveries, encounters, and conflicts with strange tribes. That mostly ended with the renunciation of polygamy, although like Grey and Whipple writers could set their plots in the past. Instead the literature of Mormon Country has been a literature of nonfiction: of journals, of family history, of the folk history of settlement, and later of natural history.[3]

Throughout the 19th century Utah's political and settlement history was one long anomaly. The endless probation that the Territory of Utah endured only stiffened resistance. It was not quite like Texas's decade as a more or less autonomous republic, but it yielded similar results. It left Utah with a rooted population and a blood-and-soil nationalism that would conflict with environmental enthusiasms in recent decades because in Utah, unlike Texas, the federal land agencies managed a large public domain; its prolonged gestation as a territory under federal oversight meant that state-sponsored conservation could intervene before Utah achieved statehood. In fact, it was Utah that inadvertently blazed the trail for that project across the country.

The distrust of the federal government still lingers in rural settings, periodically revived by presidential proclamations setting aside tracts of land as national monuments. A wilderness ideology conflicts with a story of reclamation, of making the desert bloom, that was the pith of Mormon pioneering. A passion for leaving places roadless (or removing roads) challenges a people whose creation story involved a trail across the western wilderness, and who celebrated such latter-day variants as the epic Hole-in-the-Rock expedition that traversed the slickrock country of southeastern Utah and laboriously hacked a road through a cleft of sandstone to reach Bluff. To a people who habitually acted in groups, and whose organizing conceit is a literal gathering of peoples, that canonical book of modern environmentalism, Ed Abbey's *Desert Solitaire*, would seem like self-indulgent solipsism and his plucking out of surveyor stakes for a road not merely an act of vandalism but of narrative desecration. That the merry pranksters featured in *The Monkeywrench Gang* are pursued by a local LDS bishop only highlights the contrast in worldviews and values.

Yet it leaves us with the paradox that Utah, so seemingly out of step with national norms, was, a century earlier, the model for rational settlement in the West. From Utah then came the first map of forest fires to document the relationship between agricultural settlement and fire. From Utah today has come one of the most progressive state programs for coping with exurban settlement and fire.

All times are times of transition, but for Utah (and fire) a particularly usable time for understanding falls between 1877 and 1879. In 1877 Brigham Young died, and with him the Moses that had led his persecuted people across the wilderness and then, unlike Moses, planted a civilization. His death effectively ended the heroic age of Utah pioneering. In 1878 John Wesley Powell submitted one of the founding tracts for American conservation, *Report on the Lands of the Arid Region of the United States*. In 1879 the federal surveys of the West consolidated into the U.S. Geological Survey, and the Powell Survey's *Arid Lands* report was reprinted in its most widely circulated edition.

It was a pivotal moment. As Deseret was eventually absorbed into the United States, so the Mormon commons, undergirded by divine revelation, yielded to national forests, parks, and wildlife refuges, underwritten by the revelations of science. In the *Arid Lands* report, Powell's scientific corps pulled together earth in the form of arable land, water in the guise of watersheds, air in the sense of climate, and fire to argue for a new model of settlement. It accepted as a defining reality that the Interior West was arid, that farming on the eastern model would fail, that legislation like the Homestead Acts needed to be rewritten, that a more communal arrangement had to replace ruinous laissez-faire rushes. For a model it looked to Utah. The Mormon pattern of settlement differed radically from that typical of the Interior West. Science had to replace scripture, as the state would the church, but the pattern of close settlement and a communitarian ethos should endure as ideals.

What followed is a story familiar to anyone interested in public lands, government science, and the West. The Powell Survey went national when Powell acceded to the directorship of the USGS in 1881 and began to translate his insights into political reforms. His explorations of the Colorado River had given the Major a national reputation; his tenure at the USGS gave him, for advocates of state-sponsored conservation, a prophetic one. The USGS became the "mother of bureaus" as its technocratic model underwrote many of the reforms of the Progressive Era. Powell's alter ego, W. J. McGee, was subsequently recognized by Gifford Pinchot as the "scientific brains" of the conservation movement. For many environmentalists, Powell's saga, especially as restored through its 1954 translation by Wallace Stegner into *Beyond the Hundredth Meridian*, became a species of Holy Writ. What unfolded is one of the great sagas in American environmental history.

It's worth recalling how much of that vision had emerged first in Utah. The *Arid Lands* report had linked forests, water, and fire—an inventory of critical issues in western settlement. On forests, it provided the template for later surveys of the nation's forest reserves. On water, it helped inspire the 1888 Irrigation Survey, the predecessor to the Bureau of Reclamation, and confirmed that watersheds would rank alongside timber as a motive for reserving forests. And on fire, it produced the country's first fire atlas. When Powell later defied European forestry by arguing the case for indigenous burning, he cited his experiences among the Paiute, an appeal that evolved into the light-burning controversy. A century later, echoes of those formative years still reverberated as the nation grappled with a new era of settlement and fire. The National Biological Service was modeled on Powell's USGS. Landfire selected Utah for one of its prototypes. The USGS published a map of large fires for the country in which, once again, Utah appeared as a large blotch of burns.

In brief, the modern era of wildland fire, when a fire science joined with government institutions to inform political decisions, first blossomed in the Interior West. Its significance shrank over the coming decades. Utah had almost no role in fire strategy and for more than a century had exerted virtually no influence on national thinking. No fatality fires had darkened or forced an examination of its fire scene. No founding leaders had emerged from its high plateaus or basin ranges. No informing principle of fire management, rooted in Utah experience, affected national deliberations. What geography had placed at the center of the American West, and what a proto-fire science had placed at its core, had seemingly been banished to the outer darkness of American fire.

How had such an extraordinary inversion happened? Nature contributed: the land was fire prone but not explosively so. Wildfires did not overrun crews or cities. The fires recorded by the Powell Survey had been largely set by people who would be herded into reservations or had abandoned burning for other technologies, and so had vanished. The urgency surrounding calls for fire control had faded as other regions clambered to the summit of fire crises. But so had the unusual place of Utah within the nation's larger drama. It had its own creation story, one in many ways at odds with the dominant culture. That cultural separatism was reflected in perceptions of land and its purposes and in an awkward relationship to the United States overall.

Utah had been settled by a Mormon hegira fleeing a Pharaonic America that had banished it and even declared war on it. It was settled by communities, systematically, with critical resources shared in ways antithetical to the freebooting norms of the frontier. Unlike California's saga of sudden wealth and individual freedom, Utah's story was one of patiently making the desert bloom, the collective enterprise of a beehive, and so did not enter into the country's prevailing psyche. Unlike Texas, it thrived on institutions, and unlike Texas, which has few federal lands, Utah has many, and cannot ignore the federal government, even if it remains sullenly suspicious of it. Unlike the Northern Rockies, Utah had no Big Blowup and fire control did not become a formative feature of settlement. Unlike the South, fire control was not a flashpoint of cultural antagonism. It simply didn't register. Utah had other concerns.

Today, Temple Square in Salt Lake City remains the hearth of Zion. Some 60 percent of Utah's growing population are Latter-day Saints (LDS), though of that number a significant fraction are not active. But the heritage endures. Like Texas, which has also seen a flood of immigrants yet hews to the story of the Alamo, the culture has preserved and held to the old story. And in rural areas, where most landscape fires (by definition) occur, the percentage is higher and the heritage more vigorous. It influences how Utah handles fire.

Its fires flashed on even as its fire story faded. In recent decades the fires have become larger, nastier, and spookier. The land is infested with cheatgrass and resurgent pinyon-juniper. The rebounding biotas are stocked with combustibles and houses. Settlement has revived but this time with more people than sheep; instead of tidy clusters, rural hamlets have become exurbs that propagate outward like a spreading virus. The 2007 Milford Flat fire raised the standards for Utah's largest fire. The 2013 season shocked the governor and legislature into some of the most progressive fire reforms in the country.

Much of Utah, like old Deseret, might wish to live by itself, but the fires are forcing their way in. Pioneer Day is as likely as not to have fires on the mountains as fireworks in the city park. The state's fire cycle has come full circle.

CATFIRE

Wasatch WUI

IN 1879 G. K. GILBERT, a former member of the Powell Survey, now head of the Division of the Great Basin for the U.S. Geological Survey, published the first scientific paper on earthquake prediction, for which he used the fresh scarps along the Wasatch Front. He noted that the identification of place was more important than of time. If you knew where quakes were likely, you could build accordingly and survive the tremors whenever they came. That same year the Powell Survey's *Arid Lands* report was reprinted in its most widely circulated edition, and included the country's first cartography of fire regimes. It should have led to the same conclusion for fires as that for earthquakes, that specifying place was more significant than identifying time because you could build accordingly. Instead, because so many fires were set by people, it seemed that control was possible. With control, place didn't matter.

That bias continues. Utah has expended considerable effort on earthquake engineering, as it should. A Big One could level its urban infrastructure, and perhaps leave Temple Square, along with the hierarchy of the Mormon church, in ruins. Only very recently has it considered fire, which comes year after year and progressively threatens not only city fringes and exurbs, but urban watersheds and the recreational values that are collectively a cornerstone of its 21st-century economy. Again, timing seems to matter less since the fires come so frequently. But so far they have come as a background rumble of tremors. The Big One—the sort of feral fire that has savaged the Colorado Front Range—hasn't struck.[1]

The Wasatch has been the least volatile of the Interior West's major urban fronts and is likely to remain so, particularly around Salt Lake proper. The risks are increasing, but less rapidly and less violently than elsewhere.

The reasons are many. The urban strip lies west of the mountains, so steep slopes, diurnal canyon winds, and prevailing westerlies drive fire away from houses. The border between city and legal wildland is abrupt: no rural landscape acts as either buffer or fuse, or serves as a reservoir of routine flame. The fuels are mostly grass and scrub oak—nicely combustible, but not normally eruptive. Historically, urban development was compact, a model of close settlement. Historically, too, most of the burning in the mountains followed the seasonal migration of sheep, which kept fire within sight of city folk without trampling down the slopes or through the city thoroughfares.

Times change. There are new ignitions from shooters, fireworks, railroads, ATVs, and lightning reclaiming the wild back from human hands. The urban front is spreading west into shrub and woodland, and leaping over the Wasatch summit to resort towns, and the chain of settlements especially to the south, each concentrated at a well-watered canyon mouth, are thickening and pushing into private lands against and even within public lands thanks to inholdings and checkerboard tenure. The tightly wound coil that was the plat of Zion has sprung, tossing parts everywhere. The sheep are gone, but tourists have replaced their migrations, and campgrounds have supplanted logging camps. Software campuses drive an industrial economy once based on steel. Worse, the fires have changed. What had been seasonal freshets, or resembled irruptions of flowering annuals that rose and fell with the rains, what had been a seasonal nuisance of flame and smoke, are mutating into monsters.

What is striking is how different Utah's response has been compared to its bordering states. If Nevada and Colorado looked to California for inspiration, Utah reached back to the character of its pioneering heritage and devised a program that pivots on cooperation at the lowest levels of governance. The eastern fronts of the Sierra Nevada and the Rockies built on mining: populations blew in and out like the Washoe Zephyr. Utah rose from religiously organized, close-settled agricultural communities. The Mormon village has long ago shed its role as a practical plat for living.

Instead, LDS wards make villages within suburbs and downtowns and university campuses, and the clustered village has been replaced by the 2x4 shrapnel of exurban cluster bombs.[2]

But the memory of that peculiar pioneer mix of individual agency and collective stewardship endures in its cultural DNA. When economic growth soured and slid into incoherence during the 1990s, Utahns tapped into their past to devise a version of collaborative urban planning (without calling it planning, never that) under the rubric of Envision Utah and for the surrounding countryside, updated versions of collaborative watershed protection, the Utah Partnership for Conservation and Development and the Utah Watershed Restoration Initiative. They were, in a sense, the twin legacies of Young and Powell. The sacred and the secular found common cause. When fires rumbled and menaced both communities and the land from which Utahans derived their water and livelihood, they again deep-bored into that heritage under a program titled Catastrophic Wildfire Reduction Strategy, or simply CatFire.

The 2012 and 2013 fire seasons crested what had been a rising average in fires and burned area that began, as it had over much of the West, around 1994. Utah had been spared the worst of the outbreaks. Its WUI had not been developed as aggressively or in as many awful locations as elsewhere. Its fires typically blew up and then blew out. The state forester had responsibility for all state trust lands and unincorporated lands but ran the program through agreements with the counties. The Utah Department of Natural Resources paid for big fires through an insurance fund, financed by levies on the counties, which collected taxes based on property values.

The 2007 Milford Flat fire, Utah's largest, started by lightning, burning over 363,052 acres and shutting down I-15, could be dismissed as a freak. But repeated big fire years could not. All the conditions favoring hostile firesheds were worsening; and since Utah's construction industry had evaded the worst of the Great Recession, exurbs boomed. Houses rose beyond old city limits, cheatgrass expanded beyond its former range, fires burned beyond local capacity to contain them, and expenses went beyond what the insurance fund could pay. The fires took lives and houses. Their costs were damaging. They disrupted the economy of the state.

Commissioner of Agriculture Leonard Blackham approached Governor Gary Herbert to plead for action. The response was a commission on catastrophic fire that Blackham chaired. This in itself was unremarkable; most western states had reacted similarly; commissions were a common political dodge. What was remarkable was the way in which Utah's fused national interests as codified in the National Cohesive Strategy for Wildland Fire Management with the peculiar dynamics of Utah politics left responsibility at local levels.

The final report, *Catastrophic Wildfire Reduction Strategy*, accepted all three legs of the national strategy, granting the need for resilient landscapes equivalent standing with fire-adapted communities and improved wildfire response. It transferred operational control to the state forester's office. It insisted that action had to be collaborative across jurisdictions. It asserted the need not just for fringe treatment of the WUI, but for management of the larger landscapes. It appreciated that a "broad social license" was essential if anything was to be accomplished. It counseled an urgent patience. It accepted that remediation would require decades.[3]

There remained lots of second-order observations and recommendations, and a keen awareness that traction would come through six regional committees, well anchored in local concerns, that would propose and oversee projects. The money might come from the top, but the projects would grow from the bottom. That stalled fears of top-heavy administration and black helicopters over the ridge. But giving locals control did not mean they would exercise that power or use it effectively, nor did it mean they could pay for the program. Education, close work with community wildfire preparedness plans, and pooled funding were critical, though it was not clear just where all the money would come from apart from mineral leases on state lands. For catastrophic fires the state changed its financing model from insurance to an emergency fund. The Utah legislature enacted the recommendations into law in 2015. It authorized the first funding for what became known as CatFire in 2016.

The upshot may be as striking for what it did not do as for what it did. It did not, as most western states did, put all its efforts into protecting communities: it appreciated that watersheds and surrounding landscapes were a vital part of life and livelihood and needed good fire restored (the threat of listing the sage grouse helped). It did not create a state fire service on an urban model. It did not join an interstate fire compact, on the theory that it had plenty of cooperators among its existing interagency matrix, especially the feds, even as the legislature was passing resolutions calling for the transfer of federal lands to the state. Nor could it shut down shooters and fireworks as sources of ignition (in the 2013 season some 18 fires were started by shooters around Saratoga Springs alone).

Utah's peculiarities—here used as leverage to create a unique program—make CatFire not easily transferable to other states. It's also a model yet to be field tested. Local control means local responsibility: communities have to assume more of the risk than they are accustomed to. The volunteerism of Envision Utah has had to withstand the strains that pushed urban growth at breakneck speeds. The volunteerism of CatFire would have to hold when stressed by the wildfires that crashed into those newly unplatted towns and cities.

Consider three canyons that spill down the Wasatch into Salt Lake City. Each has a distinctive settlement history, each shows a different fire regime, and each must be integrated with the others to understand the complex bonding of fire and culture along the Front.

The most famous is Emigration Canyon. Along this route the Mormon hegira ended its trek. At its mouth stands a monument commemorating Brigham Young's famous 1847 pronouncement, "This is the place." In 1961 the trail became a National Historic Landmark. A This is the Place Heritage Park, complete with pioneer village, flanks the monument. For decades the road through Emigration was a trek for sheep migrating between mountain and lowland pastures, and was frequently fired, and like similar scenes elsewhere in the state the synergy between grazing and fire overturned much of the biota. The canyon is mostly private land; the upper slopes and summits, national forest. Today Emigration Canyon is more often a trek for pilgrims and cyclists, and sheep camps have been replaced with clusters of houses. It looks like much of the changing rural West. Emigration Creek is polluted.[4]

To the north lies Red Butte Canyon. At its mouth sits Fort Douglas, established in 1862. To provide water for the fort, two reservoirs were constructed by 1875. For nearly a century a sandstone quarry marred the scene, but in 1890 (a year before the forest reserve act) Congress sought to protect the watershed by prohibiting further land sales or developments. The army began acquiring title to most of the land in the canyon, a task completed by 1909, at which point the army sealed the canyon with a gate. In 1906 it built a larger dam on Red Butte Creek, and rebuilt it between 1928 and 1930. There was some sporadic but no systematic grazing. There was very little fire. In 1969 the army transferred title to the U.S. Forest Service, which designated the canyon as a research natural area. There are few fires; the one of note, in 1988, spilled over from Emigration. Red Butte Creek is virtually pristine.[5]

To the south of Emigration runs Parley's Canyon. Since it provided a smoother passage into the Salt Lake valley, cross-continent traffic was rerouted through it, and today it serves as the corridor for I-80. As with Emigration most of the land is private, with the upper ridges and crest in national forest. It joins the urban metropolis along the Wasatch Front with pockets of small towns, mostly ski resorts, on the Wasatch Back. Two reservoirs, Mountain Dell and Little Dell, dam the upper watersheds of Parley's Creek. Without ready access from the interstate, there is little development otherwise. The densest cluster of high risk communities lies on the Wasatch backside, at places like Summit Park, where houses interweave with both native and planted conifers. As everywhere, fires cluster along routes of travel.[6]

Three slices of Utah history, three kinds of fire, three tiles that add to the Utah mosaic. Red Butte Canyon suggests a presettlement, if not prelapsarian, landscape. Surely, fires would have started historically at the mouth and worked their way up the southwest-facing, open-to-the-prevailing-winds canyon, but so far fires have not been deemed necessary, and no plans exist to "restore" them. Emigration Canyon is a relic, somewhat modernized, of 19th century settlement—its landscape not unlike the heritage park in its delta. Fuel treatment around developments is underway, but there is no effort to reintroduce fire. Parley's Canyon testifies to a late-20th-century landscape, one shaped by the automobile, with urban sprawl meeting public land, even abutting wilderness. The anomaly is that the interface is stretched out across the Wasatch. But while 10 miles separates South Salt Lake from Summit Park, that is less than 10 minutes, about as much as driving to a neighborhood grocery store. Each canyon is a symbol of Utah's heritage. Together they make a diorama of fire, past and present.

The challenge before the CatFire community is to reconcile and integrate the landscapes represented by all three canyons into a coherent program. It's a task that involves history as well as geography, that must meld culture with nature. No single of the three models will subsume the others: they each flow in time and space from the Wasatch to the cities along its front. They are bundled, and their close gathering at the symbolic heart of Zion makes this indeed the place to appreciate Utah fire.

BURNING BUSHES

UTAH HAS ALL the maladies and problem species of the Interior West because it has all the landscapes and biotas of the region. Like Caesar's Gaul, it is divided into three parts: the Basin Range, the Colorado Plateau, and bits of the Rockies to the north. It has cheatgrass, it has mountain pine beetle, and it has what the others have but that it might claim as its own red-headed stepchild, juniper, and its ecological partner, pinyon pine. It has them, in abundance and in various species, on both east and west, Colorado Plateau and Great Basin. Why it considers them a problem is itself a problem.[1]

The pinyon-juniper duopoly is as old as any flora in the region—it's not exotic to North America or to the Interior West. Its history is one of migration, part of that cavalcade of species set into motion by the upheavals of the Pleistocene and its rapid ebb into the Holocene. The pinyon especially was massaged as a food source by indigenous Americans, then massacred for fuel by immigrant Americans. The advent of the Anthropocene has set that duopoly into motion again in ways that recent settlers have regarded as problematic. Newly identified as weeds they are reclaiming places they had once lost to the axe. They are moving into new sites previously held by grasses and sagebrush. They are a woody version of the exurban migrations that have created a wildland-urban intermix, in this case, a mingling of biotas, that has proved awkward to manage. Most of the action lies in Nevada, if only because it is larger, but Nevada has so many issues with exotics that we'll let Utah claim this one.

For those trying to see a future in the flames of today, they have become burning bushes, though the Voice that speaks from them seems more Delphic than Mosaic.

For species that have been so browbeaten and so long dismissed, the range of research may seem surprisingly robust. Unfortunately most of it has been produced after the worst commodity-driven assaults on pinyon-juniper had ended. As with many places the country had to pick up the pieces after the frontier shatter-zone had passed.

In broad terms the pinyon-juniper story is one of removal and reclamation. Like the land rising thousands of years after the waters of Lakes Bonneville and Lahontan left, the two trees are still rebounding from the climatic shift that ended the Pleistocene. A great pulse of juniper swelled from 4,000 to 2,000 BP, shrank, then renewed around 1,000 BP, briefly collapsed, then pushed boldly outward 400–500 years ago. Some pinyon-juniper woodlands encountered by 19th-century explorers were still edging into fresh lands, though much more slowly than the fast-tempo pace of the new human colonization. As Richard Miller and Peter Wigand have summarized, "It is clear that re-expansion of Great Basin woodlands was just getting underway when Europeans first entered the area."[2]

Settlement repeated that scenario on a much abbreviated timescale. Vast tracts were leveled for timber and especially fuel during the mining booms; some observers, like C. S. Sargent, thought the trees were headed for outright extinction. With the axe removed, they have begun to return. People who had not seen them in precontact times now saw them reclaiming and thriving on former sites. But they also exploded over areas not known to have had them earlier. Here they began to compete with other species. They became, for many human observers, another annoying weed.

Today, the scientific consensus parses the "pinyon-juniper vegetation type" into three varieties. One is the *persistent woodland*. These are landscapes intrinsically favorable to pinyon-juniper. If the trees are removed, they will return. The second is the *pinyon-juniper savanna*. Here grasses and trees can coexist, and as circumstances change so the bias to grass or wood can shift from one to the other. The third is *wooded shrublands*. These are fundamentally shrub communities (think sagebrush steppes) in which trees can increase or decrease as climate and disturbances (think overgrazing or fire) encourage. Each variant has its own history and dynamic. This is a case where the splitters trump the lumpers.

So what is the relationship between pinyon-juniper and fire? It depends. Fires are site-specific. "Pinyon-juniper" embraces a lot of sites, which border or intermix with other species and plant assemblages, all with different dynamics. Fire must interact with other stresses such as beetles, grazing, drought, and temperature spikes. Lightning can start fires in savannas, sage steppes, and woodlands. The record of fire is sparse, especially for surface burning—too patchy to inspire conviction among many researchers. There will be a pinyon-juniper site in grasses that shows relatively frequent fire, one in big sage that shows fires every decade, one in dense woodlands that shows fires on the order of centuries. Whatever you want to find, you probably can. The danger lies in extrapolating. The data is not transcendent. Fire facts, like fires, are site dependent.

Still, some generalizations are possible. A fire regime that burns routinely will require a grassy understory to carry flame between shrubs and canopies. Given enough time the woody species can grow close enough to sustain fire during droughts and high winds, but this is likely to be over an extended chronology. The duration may be longer than the rates of migration, invasives, and climate change. The pinyon-juniper savanna can burn frequently, and the fire regime will shape the proportion of grass and trees. The pinyon-juniper shrubland will burn episodically as wetting and drying interact with an aging fuel array. The pinyon-juniper woodland will burn rarely and with stand-replacing intensity. What can seem timeless to someone driving past nominally empty landscapes may actually be a complex historical choreography of chance events.

The big disturbance of course is people. How did people interact with pinyon-juniper? How did they cope with or use fire? How should they interact today? Again, the evidence is mixed and sparse, and confused further by conflating different pinyon-juniper associations, and especially by rival beliefs, ethical and political, of what we ought to do or not do today.

What was the consequence of indigenous burning? Firsthand accounts are scarce and not easily transferred from, say, the Trumbull Range on the Arizona Strip to the San Rosa Mountains north of Winnemucca. There are written accounts throughout the 19th century of native peoples burning in the grassy savannas and grass-supporting richer soils; indigenes in Nevada surely burned as those in California did, for rabbits and small game; and there is even an account (and lithograph) of a fire drive for grasshoppers. Peter Skene Ogden describes fires along the Humboldt River. There are accounts of underburning through pinyon groves to protect trees and assist with pine nut harvests, not unlike Mediterranean farmers burning under olive groves—pinyon nuts were the great staple of the Great Basin economy. S. J. Holsinger described the practice in northern Arizona, where pinyon were the natives' "orchards." (There are no such accounts for pure juniper woodlands, which were unlikely to carry fire except after exceptionally wet winters.) And there are the celebrated observations of John Wesley Powell, then directing both the U.S. Geological Survey and the American Bureau of Ethnography, about Paiute burning in woodlands. Most of the fires were surely set for hunting in the most general sense of that word, which would include habitat maintenance and its associated foraging.

But this is less a rich stew of observations than a stone soup that can easily be dismissed as anecdotal and irrelevant across so many places and peoples. Probably the surest approach is the most general: the indigenes of the Great Basin burned as peoples of similar technology and economies did around the world. They burned lines of fire along their seasonal routes of travel, and they burned fields of fire where they paused to hunt, trap, or harvest nuts, grass seed, and tubers. In good fire years those flames would spread beyond their sites of origin. In poor years they would flicker out like guttering candles. The spotty observations in the historic literature are spot-welds to this general template. Areas that have fire scars that differ by local sites suggest the same: that the more frequent fires did not follow simply from richer fuels but from the richer resources those sites held which could be extracted by judicious burning. And of course there are always examples of injudicious burning as fire littering followed people as surely as camp vermin. Those who have an ulterior motive to find examples of widespread indigenous burning or the lack of it can.[3]

Firmer records exist for the era of American settlement, and they improve as the era unfolds. But settlement was so unmooring that it's hard to know what those records witness. Each historical snapshot can describe what conditions are like, and maybe how they differ from what existed a few years or perhaps a decade or two earlier, but they don't testify to the longer historical panorama. Even before Americans began their frenetic sprawl over the Basin, that scene was in long-wave upheaval. Those snapshots show change. But from what? and with what significance?

If there is no fixed point of observation in the natural landscape, neither is there one among the observers. People see what they are trained to see and want to see. Outside particular sites there is disagreement about whether pinyon-juniper woodlands are expanding or rebounding, and if so what that means and what might be done about it. To some observers, at some sites, pinyon and juniper were merely reclaiming their old homelands. Their spread is an ecological *Reconquista*; they are taking back what had been taken from them. Others see them as a vanguard of climate change. To many rangeland managers they have become a woody weed, even in their own landscape. To those interested in range and wildlife they were part of the black carnival of exotics that rolled into the Great Basin and refused to leave. Whether or not they were exotic, they were expansive.

A century of dedicated rangeland and Great Basin science has proved inconclusive, not just because the culture drives the science, not the science the culture, but because the disciplinary tribes of scientists can't agree on what they are recording. They follow the norms of their professions. Nearly every applied practice could point to a scientific study that appeared to justify it—even urge it. In the end the fundamental questions are not scientific at all.

What would a thumbnail historical sketch look like? The story would open with removals, full of unintended consequences, and it would close with restorations, again with unintended outcomes.

A short history over the past 150 years has a narrative arc that begins with a massive overturn mostly driven by the desire to extract ore, wood, water, and forage, and ends with attempts to return large fractions of the land to what had existed before. The clearing of the pinyon pine had the effect on Great Basin indigenes that the hunting of the bison did for Great Plains tribes. It destroyed their primary food source, which eviscerated their way of life. Then, as the woods were gone, the mines played out, and pressures relented, a revanchist pinyon-juniper duopoly began to reclaim former sites and advance onto new ones. Thanks to abusive grazing, the old checks and balances were gone, not least the perennial grasses that in former sagebrush steppes had fed patchy and pruning fires. One response, during the post–World War II era, was to deliberately convert those resurgent woodlands into pasture, often with massive land clearing operations.

Reclaimed woodlands, newly colonized woodlands—all were lumped into the same shorthand category. The chaining of pinyon-juniper was of a piece with clear-cutting, damming the last major rivers, and poisoning coyotes and fire ants. They underscored a shared assumption that applied science could ensure maximum production, a belief seemingly validated by faith in economic theory. All were criticized by the public out of a common sense of revulsion at what the scene looked like and what it cost in terms of other values. But foresters didn't on their own end chaining and clear-cutting, chemists and entomologists didn't quit using DDT, reclamation engineers didn't voluntarily stop damming. The larger culture decided that, whatever "the science" said, it wanted a different world. The sciences then moved to validate that perception.

Restoration for commercial ends then slowly yielded, during the 1970s and 1980s, to interest in restoration for ecosystem integrity and biodiversity. This could mean refusing to alter the land, or it could mean converting it to support the threatened sage grouse.

The shift from chaining off tens of thousands of acres of pinyon-juniper to cherishing old-growth juniper and celebrating the Clarke's nutcracker did not flow from research. It did not follow from new data but from a new sensibility. It radiated from a phase change in how Americans saw nature and their place within it. The culture flipped a switch that made preservation a goal, the indigenous a value, and biodiversity an axiom. The science followed.

The real issue was what to do, or what to leave alone. The chronicle of past engagements from fur trappers to range scientists did not make uplifting reading. Whether they had sought to loot or to repair, the outcomes were often wide of intentions and too often made new messes. By the new millennium restoration writ large had become a dominant theme in American environmentalism. But there were plenty of critics who viewed restoration as another guise for human meddling. They were especially alarmed that treatments proposed for other landscapes might be applied to pinyon-juniper woodlands, that the perception that pinyon-juniper was an invasive was wrongheaded, that it was "widely assumed" that fire exclusion was a cause for the propagation. Ideas mattered. If pinyon-juniper was an invasive, then it was right to roll it back. If pinyon-juniper woodlands were much denser than in presettlement times, then it made sense to thin them. If fire exclusion was a cause for invasive behavior, then it was right to kindle remedial fires. But if pinyon-juniper was self-restoring to its old habitats, then it would be wise to leave it alone.

In 2006 a workshop gathered together the major researchers under the auspices of the Colorado Forest Restoration Institute at Colorado State University, and tried to reach some consensus. The one point of agreement was that we didn't know enough—didn't know enough about presettlement conditions, didn't know what drivers had acted in what ways to shape the contemporary scene, didn't know the basic ecology. We didn't know enough to make truly science-based decisions about current and future management. The conclusion was both prudent and self-serving. We should be cautious about extrapolating from limited data. We should fund more research.[4]

But we'll never know enough. The search for a grand driver behind the changes is part of the monotheism that western science inherited from its Judeo-Christian heritage. The contributors to pinyon-juniper ecology have been many, they have changed over time, and they were driven by chance more than the chosen cause of a jealous Purpose. The story of research in the Great Basin is that science has always trailed behind events; there is no reason to think the future will be any different. As with causation, modern science will be one contributor among many. To critics it is only one more interloper imposing its own values on a land that mostly needs to be left alone. The best counsel reflected not the latest communiques from labs and field stations but ancient ethics about adapting behavior to specific circumstances and acting with moderation.

Not surprisingly, fire management reflected these uncertainties—or the conviction about those uncertainties. No formula sprouted from the findings. Critics didn't want fire as "a tool" applied as recklessly as cattle, sheep, and chaining had been in the past. Especially with cheatgrass lurking in the understory it made sense to default to fire suppression. Fire as an ecological catalyst had to interact with too many variables, both environmental and social, to be used outside particular conditions. Yet it was equally unclear that the way to avoid the abusive practices of the past was to abolish all practices.

In the end, as it always does, the conundrum came back to culture. Why was pinyon-juniper even a problem? For whom? The true problem may be with how we attempt to define the problem. With so much change in the past, and so much forecast to come, what may be breaking down in the Great Basin is the model that science advises and management applies. Science creates data, but narrative, aesthetics, and ethics create meaning. Today, we have mountains of data, libraries of knowledge, more science than ever, and perhaps the worst fire problems since *Homo erectus*. Maybe it's time to give polytheism a chance.

Like cheatgrass, pinyon and juniper are here to stay, whether or not they stay within what research has deemed their historical ranges.

PLATEAU PROVINCE

TWO STRANGER GEOMORPHIC regions would be hard to find, and to smash them together invites the geologic equivalent of a visual slap. Between the Colorado Plateau and the Basin Range an immense fault zone runs from the Wasatch Mountains into Arizona, heaving a bold cliff face several thousand feet up along its western edge. To the west, basin follows range; to the east, the summit spreads east as the lofty Colorado Plateau. It's a line etched in stone.[1]

That crisp border blurs in southern Utah. The divide is multiplied into a series of parallel faults, a delta of high plateaus and high valleys that spills and widens from north to south. Together those parallel faults resemble a kind of geologic stuttering, as though probing the crust cautiously, before making their great westerly collapse. Collectively they constitute the Plateau Province. With so much exposed rock, earth processes were visible even to innocent eyes and led to foundational studies in geomorphology and structure; for American geologists the High Plateaus are among the most celebrated geomorphic provinces in the country. So, too, their fires were so abundant that exploring geologists mapped them even if their stratigraphy stood outside the geologic timetable. Yet among American fire scientists they are among the least appreciated of fire provinces.

The High Plateaus begin at that triple junction where the Rockies, the Colorado Plateau, and Basin Range converge, then trend south, edged on the west by the Hurricane Fault and on the east by the immense warp of the Waterpocket Fold. The southerly flow of the plateaus ends in a series of vast stepped terraces known as the Grand Staircase. Usefully, the Staircase is framed by two national parks, Bryce Canyon at its top and Zion at its base. A geomorphic after tremor is visible as a bulging uplift farther to the south, also broken by north-south faults into four plateaus through which the Colorado River has excavated the Grand Canyon. The Canyon is deep because the plateaus are high. In a sense those uplifts are the last of the High Plateaus and the Canyon a more condensed and intricately carved Staircase.

In Utah the upshot is a Basin Range structure imposed on the Colorado Plateau. Naturally, there are a few geologic adjustments: plateaus instead of peaks, high valleys instead of plunging basins. The summits are elevated: Mount Dutton is 11,040 feet; Mount Marvine, 11,610; Circleville Mountain, 11,270. So are the valleys: Richfield sits 5,334 feet above sea level; Circleville, 6,066 feet; and Kanab, at the foot of the Grand Staircase, 4,970. Naturally, too, there are some ecological adaptations. The High Plateaus are sky islands that, save for low-desert salt playas and blackbrush, hold most of the life zones of the Interior West from sage steppe to subalpine forest laden with bristlecone pine. The land is semiarid. Juniper encroaches up slopes and into valleys. Cheatgrass crowds the scene to 7,000 feet and beyond.

The big difference is human history. The Basin Range defied settlement, unless one counts the transient mining camps that sprouted and wilted and the ranches that treated their grass like ore veins and also played out. The Colorado Plateau forced settlement—even in Ancestral Puebloan times—into niches, fiercely separated by impassable gorges. By contrast, the Plateau Province provided secondary corridors to the Wasatch Front for systematic colonization by Mormon pioneers committed to villages and communal life. Those towns were indelibly part of the core Mormon cultural hearth. They connected to others, they rooted. Early 20th-century conservation programs give the plateaus to the U.S. Forest Service and (after 1934) the unpatented patches of the valleys to the Grazing Service, later to become the Bureau of Land Management. Those national conservation projects imposed themselves on a landscape not only shaped by humans but over a society that considered their settlements as their historic homeland.

HIGH PLATEAUS

The High Plateaus display a fire history typical of all the compass points around them. When the Southern Utah Fuel Management Demonstration Project was conceived, the area was selected "because it is at an ecological crossroads for much of the western United States." The story of the High Plateaus—of fire plentitude, of fire poverty, of fire rekindled—is the collective story told in local idiom.[2]

The plateaus belong with the region of summer thunderstorms, which means they have plenty of natural ignitions. They have known fire since the Pleistocene inflected into the Holocene. But the modern biota came into definition with people present. Anthropogenic fires both complemented and competed with natural ones. The indigenous tribes burned as hunter-gatherers did everywhere, here focusing mostly on the montane woodlands. How much they burned in the valley outside encampments is unknown, probably some to encourage grasses and creatures that fed on grass; but their fires on the margins could move upslope when conditions allowed. Observers testified that the natives burned mostly for hunting. The natives were not many, but anthropogenic fire is most pervasive where bands, even small bands (especially small bands), move about the landscapes, kindling different sites according to a seasonal almanac. The ignitions add up; fires occur regularly where they would rarely appear under lightning alone. The prevailing biota at the time of European exploration was the fusion of climate, the high plateaus, and anthropogenic practices, of which fire was surely supreme, if not through direct application, then by its interactions with everything else people did. It was a universal catalyst.

Then that ancien régime collapsed. Settlement broke the structure of the biota and disrupted its dynamics. The project came in two antithetical patterns, one systematic and one a scramble. Mormons colonized deliberately. They explored for suitable sites, then "called" whole communities. The outcome was an instant village, closely settled at mouths of streams, with cultivated fields near town, grazing pastures further out, and the mountains subject to logging and transhumance, a pattern familiar to long-nestled European villages. The peculiar geography of the plateaus, with rims instead of ridgelines, and broad hummocky summits, made transhumance inviting and encouraged pockets of private land holdings. Meanwhile, Gentiles sprawled in their typical ways, prospecting, burning, tramp herding, passing over lands like locusts. What is disheartening is that the ecological outcome for both patterns tended to be the same. The old order crumbled.

The decline begins with the compounded shocks by which the indigenes were removed, the grasses stripped out, and fires erased. Livestock crowded out native game and ate the fine fuels that had made routine surface burning possible. Again, it was not simply their numbers, which were large, but the relentless repetition of grazing on the same sites, particularly in the spring, that culled out the native perennials. The sheep browsed on sagebrush as well, and both cattle and sheep fed on aspen. Any effort to modulate the effects were overwhelmed by migrant flocks without any village tethers. It was no afterthought that the first chief of grazing for the U.S. Forest Service, Albert F. Potter, commenced his tour with a survey of Utah reserves, and the Forest Service established the Great Basin Station, dedicated to livestock research (and its effects on watersheds), on the Wasatch Plateau.[3]

Logging also took its toll. These were not the wholesale clear-cuts that railways made possible in the North Woods or along the coastal plains by binding prime timber to national markets. They were for local consumption, to build those communities. They relied on portable mills, or where those were impossible, on ox transports for hauling, or where even ox treks were difficult, on flumes and cables to move the logs. Distance was no barrier. The prime timber—ponderosa pine, much of it old growth—was soon cut out. Timber was hauled to Cedar City down tracks along Coal Creek Canyon and to St. George from Mount Trumbull 70 miles away. Cables carried logs from the tablelands above Zion Canyon to the settlements along the Virgin River. Pinyon and juniper were cut for fuelwood, especially for mines. Then loggers started on the subalpine conifers.

The High Plateaus were accustomed to disturbances. They routinely knew blowdowns, ice storms, insect infestations, floods, and of course fire. But the new onslaught happened so quickly, on such a scale, so much in sync with one another, so outside their evolutionary experience that recovery was slow. By the onset of the 20th century, what of the uplands had not been homesteaded or purchased, was committed to forest reserves. That brought some order to grazing and logging but it only aggravated the absence of fire. The U.S. Forest Service made fire protection, which really meant fire exclusion so far as possible, a mission.

The tipping point varies by place. It came to most of the Plateau Province by the 1870s. Before then, across much of the region, landscape fires continued to burn with the rhythms of seasons. Then those fires flickered and faded away. The record of fire-scarred trees complements that of repeat photography. Fire vanishes. The big losers are the native grasses. The winners are woody vegetation: shrubs, scrub oak, juniper, white fir, pine and Douglas fir reproduction. With one revealing exception.[4]

Every American fire province has its signature species. The Southeast has the longleaf pine; the Northern Rockies, lodgepole; the Southwest, ponderosa; the Great Basin, sagebrush; the Great Plains, tallgrass prairie; Southern California, chaparral. Each province has what it judges to be a species problem: it has too much of what it doesn't want, typically invasives, and too little of what it does want, normally natives. The Great Basin has too much cheatgrass and not enough sagebrush. The Southeast has too much hardwood understory and not enough longleaf savanna.

For the Plateau Province the signature species is aspen. In some ways it's an odd choice. Aspen is not a big-haul timber tree, the region is not celebrated (except locally) for its fall colors, and while it serves as critical browse for deer and elk, no one has taxidermied an aspen to hang alongside trophy elk heads. Aspen woods are not celebrated in art and literature and local lore. Aesthetically, they have delighted rather than inspired. The best known Mormon novel, *The Giant Joshua*, spoke to colonizing in the southern Great Basin and was named after the Mohave Desert's Joshua tree. Aspen were not prize habitat for endangered spotted owls or scrub jays. The aspen was just there. It had always been there, part of the biotic wallpaper, as unchanging as the lava-crusted rims, as little fretted over as sage. At the time of settlement, it was probably the dominant tree. And then it wasn't. By the late 20th century, they were no longer supreme. To some observers up to 60 percent of Utah's aspens were gone or going and not regenerating.[5]

The plateau aspen were losing ground on all fronts. Above, subalpine conifers were encroaching and were primed to overtop decadent stands. Below, sagebrush and juniper were reclaiming lands once rife with aspen. Failure to sucker, or suckering cropped to soil by browsing, meant that aspen lay latent in the soil, like a cellulose seed bank, while the surface was rebuilt with other species. Even the middle stands were thinning and aging, ready for pruning and rejuvenation. The prevailing sense was that renewal had to occur on a large scale. Prior to industrial settlement, this meant fire. The dead leaves could burn in spring and fall. Invading conifers could stoke stand-leveling burns on the order of decades or centuries. In 1925 Frederick Baker observed that, although "all the evidences of past fires in aspen are plentiful, and extensive fires have run through the type within living memory of living men, they are extremely rare at the present time." Fifty years later those fires were forgotten, and aspen with them.[6]

Aspen landscapes got little attention from fire researchers who focused on species famous for surface burning or for explosive crown fires. It seemed an interim species, a placeholder for the hardier conifers. It was more likely to stop fires than sustain them—aspen greenbelts were often proposed as fuelbreaks. Its green leaves wouldn't burn. Its dead leaves burned, it seemed, with no more effect than scurrying voles. They were unlikely to threaten exurbs. They did not stockpile combustibles, so were not amenable to the prevailing fire-research methodologies. They were not a fire problem, so were not a topic for fire research. Yet they were becoming an ecological concern, and their nominal fire history did not make sense when probed.

Gradually, by the 1980s, attention turned to their rapid decline. They were valued browse, which attracted wildlife biologists and hunters. They were embedded in cycles of conifer fire, and it occurred to some observers that they were more than the ecological equivalent of junk DNA in the dynamics of those forests. They even found a cultural bond as the realization grew that aspen did not, in the plateaus, reproduce much from seed but by repeated suckering, over centuries, over millennia, from buried roots. Whole hillsides were single individuals displayed as cohorts. They had been renewing themselves for thousands of years before the advent of the Holocene. They might be among the most ancient living things in North America, older by a factor of five to eight compared to redwoods, and perhaps double the age of the oldest bristlecones. They were becoming living ruins, like the tumbled terraces and stone houses of the Anasazi. One giant clone in the Fishlake National Forest, known as Pando (from the Latin, "I spread") or the Trembling Giant (after its "quaking" aspen common name) has 47,000 stems across 106 acres. At an estimated 13 million pounds, it may be the heaviest organism on Earth. At 80,000 years its roots may be the oldest living thing. In 2006 the U.S. Postal Service issued a commemorative stamp in its honor within its natural wonders series. Aspen, in short, are the old-growth trees of the Plateau Province.[7]

Or not. As with every other species of interest in the Interior West, there is scant scientific consensus about aspen and fire. There are sites where seeding seems vigorous and where encroachment may be more apparent than real. The role of fire is, inevitably, controversial, though there can be little doubt that the abrupt recession of fire has contributed to the general senescence of the species as a dominant forest.

New data and new interpretations come and go like El Niños. Behind the controversies, however, lies the recognition that the issue is not simply ecological but historical, that the massive changes of the Holocene make a reference point impossible, that quaking aspen has been on the High Plateaus for perhaps 80,000 years and modern science for less than 150, and that history, whether natural or human, has its own logic grounded in narrative.[8]

The facts may be unsettled, as they are with regard to aspen, but there can be little doubt that science and narrative make, at best, a shotgun marriage. Narrative is teleological; science is defiantly not. They are two forms of understanding that just do different things. Paradoxically, the hard data of science is so much sand. Build a narrative on it and the anchors of your arc may well be swept away with the next downpour of discovery. Better to build instead on the hard pilings of aesthetics, or the aggregated concrete of philosophy, ethics, and literature. The harder fact, however, is that this uncertainty applies not only to what we know but to what we do. The fire community's appeal to science-based management has not led to firm principles on which to base action but, where successful, to programs of continual adaptations, the kind of mutual trial and error that is the basis for traditional ecological knowledge, this time coded into algorithms and simulations rather than stories and ritual. Much as science filled the void left by the loss of traditional lore, so modern narrative had to replace the old myths and legends.

The important outcome is that High Plateau aspen entered into the realm of biodiversity and socially sensitive species. It might well be a relic from the Pleistocene like giant sequoias and condors. Much as sagebrush, juniper, and cheatgrass, aspen began to matter culturally in ways it hadn't before, which pushed its significance for the fire community beyond models of fire behavior. Aspen might not be important to fire, but fire was important to aspen.

"Settlement" makes a convenient prime mover for what unhinged the High Plateaus. But that motive power had lots of auxiliary engines and drive belts. Removing the indigenes took away the keystone species, affecting not only fire but wildlife once hunted and plants previously gathered; the knock-on effects rippled throughout the biota. Cattle and sheep supplemented elk and deer, and then supplanted them, before the bovids returned, and the population of ungulates exploded. The montane forests fell to the axe. Streams were redirected to irrigation canals. These were the visible agents of a macrobiological change; the microbiology has hardly been explored. The upshot, however, is that a new order settled on the land.

There are assorted indices of these changes. Fire-scarred trees track the loss of fires. Repeat photographs show similar changes, though most early photos date well after the first shock of settlement, from the 1870s or even young 1900s. There were no formal exclosures erected by which to measure the consequences of settlement against leaving the land alone. Forest reserves and parks did not appear until the 20th century, and these regulated, not abolished, axe and hoof.

Except at Zion Canyon, carving its deep channels through the Staircase. Here the geological faults and fissures that accompanied the creation of the plateau province invited erosion that left the southern tips of the plateaus as peninsulas and isolated mesas. They became ecological islands. Especially with regard to fire they are not true experimental controls since most landscapes are part of a pyric catchment basin in which fire flows into and across sites, and the firesheds for islands can only burn from fires that start on them. But those mesas are the closest proxy history offers.

The sites that settlement could reach with livestock and axe underwent the same transformation as their surrounding countryside. Horse Pasture Mesa (as its name suggests) was grazed. Cable Mountain (as its name notes) was logged, and the timber lowered to the canyon floor by an elaborate wire cable system. But other mesas remained inaccessible, and some, like the Great White Throne, were not even climbed until the mid-20th century. (Even so, their fire history was distorted because fires on them were suppressed when helicopters became available to carry crews.) Those islands display a biota tweaked from that on the mainland. They have a different fire history.

Horse Pasture Plateau, really a peninsula, was described as "fully grazed" in 1875, lightly logged from 1905 to 1922, then moderately logged in its northern stretches in the 1950s. By 1890 fires that had previously burned on a rhythm of four to seven years, and often one to three, were disappearing, and a grassy savanna was replaced by a woody understory. When grazing pressure was lifted by the park, active suppression kept fires out. So far, so familiar. By contrast, Church Mesa remained isolated and uninhabited, though it held a population of mule deer and cougars. From the mid-18th century to the mid-20th, it experienced four fires, two of them aligning with fire years on Horse Pasture Plateau, and two after. Here people did not eliminate fires, directly or indirectly, but neither did they start fires. The biota endured as a ponderosa pine savanna.[9]

Clearly, natural fire had contributed a foundational baseline load of burning. But the recorded fire history is one of people adding to, subtracting from, and rearranging that inheritance. In the Plateau Province, as throughout the Interior West, or the country for that matter, fire history without human history is science fiction. As researchers behind the Zion study note, "Without further human intervention, a return to either pristine vegetation or fire regime is probably impossible. . . ." It's a remarkable admission for its place, a national park, and for its time, the high-water mark of the American wilderness movement.[10]

MOUNTAIN SENTINELS

The Grand Staircase ends the southerly flow of the Plateau Province. Two mountains flank them to the east and west, like lion statues guarding a stairwell. Both are laccoliths—an igneous feature typical of the Colorado Plateau in which a volcano erupts far beneath the surface, and then becomes visible through massive erosion. But whatever they share as geologic features, they differ in ecological and human history. The Pine

Valley Mountains on the southwest corner of the Staircase loom over an ancient corridor of traffic and settlement: they are visible, long worked, and integrated into their surroundings. The Henry Mountains are remote, disconnected, the last mountain range to be explored in the continental United States. Why would their fire histories look so similar?

The Pine Valley Mountains—a massive massif—lie along the west border of the Hurricane Fault. They loom over a geologic briar patch of rims, mountains, cliffs, basins, mesas, gorges, a landscape of bright colors and heaving perspectives. Contrast seems too mild a concept for the scene: the elevation difference between Signal Peak at 10,365 feet and St. George is 7,505 feet. There seems no place to rest the eye, no place with a level horizon. Nothing seems plumb, everything seems out of round. That applies to more than rocks.

The geologic activity is matched by human history. The Old Spanish Trail from the Great Salt Lake to southern California flows around the massif; Mormon settlements string like beads along it. The 19th-century emigrant route through Mountain Meadows passes around its west slopes. The 20th-century thoroughfare that joins Salt Lake City to Los Angeles, I-15, runs along its eastern foothills. Both join at the town of St. George, the southern capital of Mormon Deseret.

It's surprising that, for something so prominent, no formal fire history has been written. The fundamentals are clear, however, and because of its historic interest, Mountain Meadows has long been tracked for its shift from premier grazing grounds to weedy steppe. Begin with a central fact that the Pine Valley Mountains are relatively well watered. They lie a long way from the rainshadow of the Sierra Nevada; they capture the first moisture that makes it over the fetch of the Mohave Desert. The interior valley that gives the mountain its name is, compared to the Great Basin, a Shangri-La. There is plenty to grow, which means there is also plenty to burn.[11]

The mountains show a fire history comparable to other sky islands of the Great Basin, though strengthened by proximity to what for the region is a major cluster of settlement. They were grazed heavily. They were logged early. (There are even cohorts of big-tree stumps that locals have come to regard as historic legacy.) The fires went. Lightning still kindled starts, but they were suppressed, and the old pattern by which fires set in the montane forests of the foothills could race uphill ceased. As it does throughout the region, the chronicle of burns stops.

Then, as again throughout the region, fire returns as big burns. Two dominate recent times, both started well up the slopes by lightning. The 2009 Mill Flat fire burned in wilderness and was mostly monitored before it blew up. The 2016 Saddle fire, with Mill Flat still festering in the public mind, was more actively handled. Their pairing illustrates the fascinating phase change to managed wildfire that is refashioning wildland fire in the West. Both bind the region to a national narrative.

Transitions are often troubled. Geographically, the Mill Flat fire, kindled by lightning on July 25, 2009, started within the Pine Valley Mountains Wilderness of the Dixie National Forest. Bureaucratically, it sat within a switch box that toggled from a policy that gave fires a single character and left wilderness fires, where possible, to be largely monitored to one in which any fire might have several purposes and be actively managed in various ways. New implementation guidelines were to be stiffened by use of a wildland fire decision support system. All this, however, proceeded in syncopation with the fire. During the critical phase, incident management teams were also in rapid transition.

What mattered most, though, was that the Mill Flat fire itself was morphing from a shuffling surface fire that had sloughed over the ridgeline into a blowup that raced down ravines into the old community of New Harmony. A month after it started, the fire had reached 971 acres. By August 28 it had boosted to 1,295 acres. Then what could go wrong did go wrong, and what had seemed adequate by way of forecasts, protective fuelbreaks, and protocols was lost in smoke. On the 29th it quadrupled to 5,253, as managers ordered a Type 3 incident management team, and then a Type 1 team; the next day's spillovers added another 50 percent increase to 7,641 acres; and the day following the transition to a Type 1 team was complete. Hardly a megafire, but size has little to do with cultural impact. The critical fact was that it blasted into the hamlet of New Harmony, burned six houses and damaged many other structures, and forced an evacuation. (Disclosure: my wife's octogenarian parents were among those evacuated.) All this happened on a mountain that looms over the countryside, and smoke uneasily visible from the region's two major towns, Cedar City and St. George.[12]

When the 2016 Saddle fire broke out on June 13, again in the Pine Valley Mountains Wilderness, that bureaucratic pivot in policy was complete. An initial reconnaissance pointed to many obstacles; its remoteness and ruggedness, competing fires with greater urgency, its wilderness setting. The fire steadily doubled. By the 16th it was 103 acres; on the 17th, 163; on the 19th, 318. The Dixie went big and long. Keeping its Type 3 team, under pressure from the regional office, it ordered a National Incident Management Organization team and reckoned the expense was merited because the fire could threaten the community of Pine Valley and the forest's recreational complex. By the 22nd the incident management team had worked out a box, begun assessing and prepping the community, and cleaned out fuelbreaks that traced the border of the national forest. By July 1 the Saddle fire had reached 1,540 acres and was 42 percent contained. Some monsoon rains dampened both spread and burnouts, but allowed for patches of direct control. On July 6 NIMO devolved to the local Color Country T3 Team 1. The fire was 1,647 acres. Its cost reached a heart-stopping $10,254,400. Before it was declared out the Saddle fire burned over 2,299 acres.[13]

Lessons learned, and lessons still a-learning. Monitoring fires passively works, until it doesn't, because when a fire appears ready to boil over, it is too late to put it back in the kettle, and the longer a fire lingers on the land during fire season, the more likely it is to draw the cards for a full house. Better to box it from the beginning, provide critical protection, and burn it out so that risk and smoke don't hang for weeks over the landscape. Boxing and burning is a hybrid strategy, part suppression, part prescribed burning. It allows fire officers to do under emergency conditions what needs to be done. But so is protecting communities. The emergency preparations to shield houses can do under the aegis of the wildfire what was too costly, onerous, or controversial to do beforehand.

The transition to managed wildfire as a foundational strategy is still underway. A monster mistake, a flip in national politics, or a legal challenge could unravel it. But at the Pine Valley Mountains there appears to be no Plan B. Suppression is self-defeating, prescribed fire is too precious and encumbered, mechanical treatments are too expensive, cumbersome, and controversial. A managerial mashup seems the only alternative.

The Pine Valley Mountains are a massif; the Henry Mountains, an archipelago of peaks rudely ranging from north to south. Hard to reach, not on the road to anywhere, they are a destination site only for geologists and hunters. The Henrys are the type locale for laccoliths and the inspiration for one of the classic studies in American geology (G. K. Gilbert's 1877 *Report on the Geology of the Henry Mountains*). They also hold trophy mule deer, and by a quirk of history, a herd of bison. But they are not known for fire.

Still, repeat photography and fire-scar chronicles have been assembled. They play out the same rough scenario that characterizes the Plateau Province overall. The fires began disappearing during the 1870s, about the time the native Paiutes departed. Some logging with portable sawmills occurred to build Hanksville and other settlements. But livestock moved in during the late 19th century and reached epidemic proportions in the early decades of the 20th. The grasses went. The fires never returned. Lightning kindles perhaps two a year. Throughout the 20th century, routine burning was unknown, and until the 21st century, big fires were, at best, a legend.[14]

At the Pine Valley Mountains the fires left when livestock and settlers arrived. At the Henry Mountains they receded when the indigenes departed. Then overgrazing took out the grass and prevented a return. Later in the chronicle, fire suppression knocked down what starts did occur. Woody vegetation thickened, particularly the pinyon-juniper woodlands along the lower flanks. In 2003 the Lonesome Beaver, started by lightning, and the Bullfrog fire, kindled by people, swept over 34,000 acres, much in pinyon-juniper. The Bullfrog burn moved into stands of big ponderosa with unprecedented severity.

What was true for the Henrys' geologic history is thus true for their pyric history. The present scene is the outcome of erosion, in this case, the loss of anthropogenic fire, the scene then confirmed by later erosions of fuels and direct suppression. The mountains are high—Mount Ellen at 11,527 feet rises higher than the summit of the Pine Valley Mountains. But the Henrys rest within the rain shadow of the High Plateaus, and they are wildly broken country, full of ravines, gorges, hogbacks, and sculpted peaks. A fire that starts in one place will not easily spread across any one of those sky islands, much less several, though surely there were times when wet winters boosted grasses, lightning found the right snag, and droughty summers left those fires to creep and sweep through mixed terrain over months.

Yet the evidence suggests that the old fire record had to be primarily the result of frequent, recurring ignitions, inflaming pixel by pixel. The Pine Valley Mountains are sufficiently verdant that fires can burn for weeks, sometimes seasons. The Henry Mountains are so isolated, broken, and comparatively arid that they can burn collectively only through many, scattered starts. If Pine Valley testifies to the power of axe and hoof to break old fire regimes, the Henrys bear witness to the power of indigenous burning to put fire into places that would not, under purely natural conditions, have it.

Viewed up close and personal, the tangle of ravines, mesas, arroyos, ridges, and peaks that constitute the Henrys can overwhelm with detail and complexity. Better to view the range from afar to comprehend its structure. From the rims of the High Plateaus, you can see a hundred miles. The geology of aeons spreads before you like a tablecloth. So it may prove with fire history. You need the long view. But that is what the High Plateaus so boldly offer.

LONG VIEWS, BIG NARRATIVES

Turn that long view into narrative. A quick survey of the Plateau Province—the equivalent of the kind of interpretive exhibits posted so often at overlooks—suggests a familiar tale, a story of fire vigorous, then of fire suppressed, and finally of fire both haltingly restored and virulently gone rogue. It's a standard chronicle of fire for the Interior West, true from Canada to Mexico. But a chronicle only becomes a narrative when it has a theme, a narrative arc, and its vision has moral spectacles. What narrative speaks on the Plateau Province?

The default narrative here, as generally, is what environmental historians like to call declensionist. It's an ancient trope that speaks to an age of gold in the past that has decayed into today's lead. It's the story of a garden that degrades into dust and weeds. The template, a morality tale, long predates Modernism's predilection for irony, but in western fire an old template and modern taste have bonded readily. The story of American settlement as a narrative of progress becomes, instead, a narrative of decline. The contributing causes migrate among commentators, but capitalism, ignorance, greed, arrogance, racism, and cultural solipsism rank high in most listings.

The High Plateaus complicate that story, as they do the expectations of fire ecology. There are critical species, not charismatic, that matter, like aspen. And there is a settlement pattern quite at variance with western norms. Wallace Stegner once distinguished between western boomers and western nesters. The boomers raced from one gold rush to another, looking for a quick placer of money, leaving messes for others to pick up. It's a world of the unattached young, of fast money and footloose values. The nesters built functioning communities full of families, schools, churches, and attachment to the land. The boomers epitomize the declensionist narrative with embarrassing ease, and in the Interior West they are most obvious in Nevada. The nesters seemingly announce the more responsible alternative, and they tend to cluster in Utah.

The Plateau Province had both boomers and nesters, but the nesters—Mormon communities systematically called and abided—were the core. They seem to stand outside the received narrative, a peculiar people with a peculiar environmental history that, like their religion, took many pieces from their times and assembled them into a new order. They were looking for sanctuary, they were willing to settle places most Americans considered valueless, they organized their use of grass, forest, and water on broadly communitarian grounds. Instead of channeling streams to wash whole hillsides into mining sluices, they carried the water to fields of alfalfa and corn. Instead of stripping ore veins and loosing maverick herds, they planted sugar beets and Lombard poplars. Of course Utah was never hermetically sealed. Forty-niner gold and Gentile markets saved Deseret from a downward spiraling autarky; there was a long probation during which Gentiles roamed the landscape before statehood arrived. Utah's narrative reflects those exchanges, infections, and revisions. But until the postwar's city-states appeared, those Mormon settlers created the only rooted landscape in the Interior West and a parable of what pioneering might have been.

That's why the historic record can leave a queasy feeling, because the environmental outcomes of Gentile and Mormon were uncomfortably similar. The landscape of both unraveled. Cheatgrass infests the High Plateaus as it does the Great Basin. Juniper spreads across Zion as well as Mammon. Perhaps it was simply the shock wave of new species and new forms of usufruct, of maximizing production according to precepts devised elsewhere, of an invasive culture that came from elsewhere and tried to recreate their old landscapes, foods, and social life in new settings. Maybe we are only seeing the yet-unfolding story in midclimax, an experiment that has a long time yet to run, with the narrative stopped in midarc.

Or it may be the true narrative is a mashup. It's neither ever upward or always downward, but like so much of life in the Great Basin an endless succession of shocks, migrations, and adaptations. The progressive and declensionist narratives have the elevated tone of an epic. They are to written history what sequoia and longleaf are to the literature of fire ecology. The narrative appropriate to Utah is a more humble, practical one, the equivalent of growing sugar beets instead of panning for gold. It's to historiography what aspen is to the High Plateaus.

Because it does not arise spontaneously from data, narrative is unstable, and for postmodernists, inherently suspect. It imposes order and moral meaning that can shift with whatever social, cultural, or political perspective animates the teller. Yet data is just data. It acquires meaning only when it is framed by aesthetic sensibility, ethical codes, or, yes, perhaps more commonly and inevitably, by narrative. If a dominant narrative no longer makes sense, if its prescriptions of the world make life harder for believers, then a new narrative will replace it.

When Wallace Stegner published *Mormon Country* in 1942, he called Utah "the last of the sticks," yet concluded it was also "a country that breeds the Impossibles." In the decades that followed, the value of the landscape as scenery and empty space remade it into a Mecca for tourists. Utah came close to the navel of a wilderness movement whose battlegrounds—Dinosaur, Glen Canyon, Grand Canyon—ran through its patch of the Colorado Plateau. The narrative of postwar American environmentalism anchors one of its spans in the Plateau Province.[15]

So, too, for a long time, for the American fire community, the area lay in the sticks of national pyrogeography. The restoration of fire here, as elsewhere, will not be simple, or rational, or made easily understandable by our prevailing narratives of progress or declension. Progress ends in irony.

Declension fails to inspire. The future will have to write a different story that may take the form of a narrative mashup, or of working narratives rather than grand, epic ones. In this it may resemble the state of wildland fire management, which cannot survive by suppression or restoration alone but by awkward minglings of the two in ways unexpected. Done right, box-and-burn is a hybrid fire, half suppression, half prescribed burn. We are neither running from the wild nor reaching toward the tame but passing between them both. What, for fire history, will be the thematic analogue of box-and-burn?

We might be surprised by what happens. A future writer with Stegner's talents and empathy may again find that this is a country that still breeds the Impossibles.

OUTLIER

MESA NEGRA

BETWEEN 2000 AND 2003, on the imposing cuesta overlooking the Mancos River, wildfires burned 39,178 acres, or 70 percent of what had burned over the past century. The 2000 Bircher fire alone scorched 23,607 acres. Those burns amounted to 75 percent of the total land under management. The fires menaced structures. They threatened prized cultural and biological assets. They prompted mass evacuations over a single, lengthy egress road. They defied normal suppression methods.

So far, so typical. These were the years in which a millennial drought blasted the greater Southwest, in which fires lit and fires fought escaped control, and in which the American fire community struggled to double down on efforts to restore fire and rekindle ecosystem health. A casual observer, even a knowledgeable pundit, might regard the great cuesta as synecdoche for the American West. The power of abstraction might toss this fire scene into the growing lump of lands with too much bad fire and too little good.

But a fire is a specific event in a particular place, and here the consequences of abstracting and amalgamating disguise what makes the massif trending into the Mancos significant. The grand cuesta is best known as Mesa Verde, and it holds one of the largest and most significant collections of Ancestral Puebloan ruins anywhere. Multifloor structures like Cliff Palace and Far View Tower are its WUI. Line construction must thread among lithic mounds, surface pit houses, and terraced fields—the analogues of threatened species and sensitive habitats. The combustibles powering most mesa conflagrations are pinyon-juniper woodlands, many of the landscape patches ancient—biotic relics as valued by ecologists as the stone walls and pot shards are by archaeologists. The usual treatments, thinning and prescribed burning, will likely damage old-growth trees and invite regime-changing invasives such as cheatgrass. The only management response is suppression, and suppression of a sort that must plan for the once-a-decade or two blowup.

In fire as in other matters, Mesa Verde harks back to an earlier era, and it reminds us that fire management varies place by place, time by time. Fire regimes, like civilizations, can rise and fall.[1]

Mesa Verde is not known as a fire park—not a place that routinely burns, that hosts a distinctive fire culture or fire species like giant sequoia or longleaf pine, that year by year commands national attention for its fire scene. The dominant vegetation is notorious for either not burning or blowing up. Most fires are one-day wonders, or as the saying goes, they either burn a tree or a mesa. Partly that reflects efficient fire suppression that has eliminated the middle class of outbreaks. The cuesta is, in fact, littered with the legacy of 10- to 100-acre patches of previous burns. The blowups of the modern era run 2,500–4,500 acres. At 23,607 the Bircher burn stands alone.[2]

It's hard to support a vital program, largely staffed by seasonals, when most fire personnel may never see but a small fire or two during their tenure. Human-caused fires are nil (access to the backcountry is prohibited); ignition depends on lightning's lottery. When the park was established in 1906, fire was not perceived as a problem. Fire records only begin in 1926. An effective infrastructure commenced with the Civilian Conservation Corps (CCC), whose buildings still house the park fire cache and fire management office. That year marked the first of the modern blowups, two fires that splashed over from the Ute Mountain Reservation border. In 1959 the Morefield fire burned 2,500 acres.

Then the 1972 Moccasin Mesa fire scoured out 2,680 acres while a phalanx of bulldozers cut a fireline six blades wide. Whatever damage the fire might have wrought was negligible compared to dozer lines that ran relentlessly through surface ruins. The Park Service moved to change fireline practice to avoid a repeat. (During the 1977 La Mesa fire in Bandelier National Monument, archaeologists accompanied line scouts in what has become standard protocol.) But the fire itself fell within historic bounds. Despite occurring at the climax of a revolution in fire policy, the issue was never about fire's possible restoration—it wasn't wanted—but the style of suppression. In 1996 the Chapin 5 fire seemed to mark a phase change in the character of landscape burns. Then came the Bircher and Pony fires of 2000, the Long Mesa fire of 2002, the Balcony House Complex in 2003. The blowups were beginning to burn into one another. Suppression was beefed up with better aerial attack. If the fire scene thereafter calmed, it was partly from exhaustion.

These were the kind of high-publicity burns that set many parks and forests to re-evaluate and repurpose their programs. Places with wilderness or undeveloped backcountry began to back off and allow fire more room. Places suffering from ecosystem degradation sought to restore fire by prescription burning. Places hamstrung, and hammered, by the presence of urban and exurban sprawl sought to shift the burden of protection from land management agencies to communities and homeowners. And a few places, most spectacularly Southern California, were simply, admittedly, unmanageable and had no future prospect of a sustainable fire program. They could only bulk up their suppression capabilities.

Mesa Verde was too small to tolerate free-burning fire, which in any event would ramble for a day or a few (the Bircher blowup lasted three days). It was hobbled in prescribed burning because surface burns damaged its fire-sensitive woodlands and invited invasives. It could not negotiate with county commissioners and homeowners because they had decamped in the 13th century. Any mechanical fuel treatments met with resistance: locals didn't like it because it exposed their residences to public view, archaeologists didn't like it because it exposed surface sites and might encourage visitors to walk through them, and environmentalists didn't like it because the disturbed sites could siphon in invasives and might damage old-growth pinyon and especially juniper. Still, some clearing and cleanup around developed areas and exhibits occurred during the 1990s. The treated areas survived the blowups while everything around them burned to white sticks and ash.[3]

Meanwhile, the firescape was worsening. The cuesta dipped southward and was gouged into north-south trending flukes and mesas, which aligned nicely with prevailing winds. In extreme times they became raceways and flumes to carry flame. And although a pinyon-juniper woodland might burn in stand-replacing immolation every 400 years or so, and the petran chaparral characteristic of the northern cuesta every 100 years, the lack of middle-range fires was allowing a more homogeneous cover to blanket the rims and canyons. The upshot is that the Mesa Verde fire program is left with minimal treatments to counter their maximum fires. They have to rely on rapid detection and initial attack. A helitack crew has the capability to land within a quarter mile of any start. A single-engine air tanker is stationed at Cortez. An air tanker base lies ready at Durango.[4]

Even so, minimum impact suppression in a World Heritage Site might require an on-ground archaeologist to advise on tactics. So dense is the record of those earlier peoples, so thronged is the surface with lithics, structures, shards, and even refuse pits, that anything a fire crew might do will impact something of apparent cultural significance. Kick a stone and it might be part of a pit house. Scrape a shovel line through dirt and you will probably disturb a pottery shard, an arrowhead, or a bone scraper. It's as though, in a typical park, every plant and animal was a threatened or endangered species. Given Mesa Verde's core mission, one can understand why the park might wish the fire crew and all its apparatus away. But banishing a fire crew will not banish the fires. Catching snag fires before they can spread is the least intrusive approach.

Suppression, however, is only one of a series of fire-archaeology encounters. Cultural resource management, too, has its good fires and bad fires. As happened at Custer Battlefield in 1984, fires of this severity can both damage exposed relics and unveil new ones. After the Chapin 5 fire, survey teams discovered 372 previously unknown cultural sites—this in one of the most intensively studied archaeological sanctuaries in the world. It is estimated that another 2,000 sites may emerge from resurveys of the Bircher and Pony fires. Some damage to existing sites did occur, mostly by spalling, but fires had undoubtedly rolled over these sites many times before the park began to document them.[5]

Modern Mesa Verdeans put their fire into machines. Ancient ones put theirs into hearths. And there, in the layered ash, is where the record of landscape fire is recorded as the composition of species used for fuelwood shifts over time. At first it was pine and juniper, and then oak and cliffrose, and then whatever shrubs and combustibles could be found. Undoubtedly people used fire for hunting and foraging outside their main settlements, but as those settlements thickened, the reach of firewood harvesting, and then of scavenging, expanded. The land was being picked clean. Landscape fires thinned, and then shrank into the hearths of kivas and kilns. In the later American Southwest, landscape flames disappeared down the gullets of sheep and cattle. In Anasazi times they vanished into domestic appliances.[6]

The scene recalls the story of early colonists in New England who were asked by the natives if they had left their old homes because they had exhausted their reserves of firewood. In swidden agriculture the village would be moved routinely, not only to let the fields go fallow for a few years (or decades) but because the fuelwood needed for domestic uses was exhausted. The great dwellings at Mesa Verde, however, could not be packed up like tepees or rebuilt like wickiups. They rose higher and wider, and they took more and more fire out of the land and put it inside their stone edifices.

There are ruins in Long Mesa, known as Fire Temple, where fire was believed continuously maintained, likely a public utility from where new fires could be kindled that may over time have become a sacred site as well. But the deeper story may be that it indirectly speaks to the narrative of landscape fire because those once-ranging fires that searched out fresh fuels were here domesticated and sustained by bringing fuels to them. Unlike landscape fires that burned sporadically and by seasons, this one burned continuously.

Only when that combustion pressure lifted, only after the population abruptly left, did free-burning flames begin to reclaim the cuesta. Eerily, it began when the abandoned villages were fired, apparently from the inside, whether as part of a ritual cleansing by those leaving or as fire-and-sword destruction by those who replaced them. At 700–800 years the oldest junipers known at Mesa Verde may very well date to the time of abandonment. More date from the 17th century when regional replacement populations crashed following the introduction of Old World diseases. The rest took root during the more developed era of heavy grazing that began in the late 19th and early 20th centuries. As people and livestock left, however, the woods not only aged but thickened, and fires burned out patches from a single tree to a few dozen acres. Then, before landscape fire could reestablish a quasi-natural regime, modern fire protection arrived and big burns defined the prevailing scene.[7]

The past is indeed a foreign country. But so, too often, is the present. Geologists are fond of asserting that the present is the key to the past. But for historians, and any strays from the fire community fascinated by Mesa Verde, it's the past that is key to the present because along with Spruce Tree House and Long House, along with one of the great concentrations of archaeological wealth in North America, field schools that trained several generations of archaeologists, and contributions to America's understanding of its *longue durée* past, the Mesa Verdeans bequeathed today's park fire policy and protocols.

The people who abandoned Mesa Verde left two legacies. One is the built landscape that attracts tourists, archaeologists, and the National Park Service. The other is the recovered natural landscape, which led to the modern firescape. The two continue to interact, but moderns are not allowed to do with their surroundings what the Ancestral Puebloans did, which is to thin and clean up so thoroughly that surface fires became implausible and crown fires impossible. By living on the land, the Verdeans used it, and by using it they kept wildfires at bay. Modern Verdeans don't live on that land, don't use it except as a source of information, and can't manipulate or transfer its fire regimes. Despite their industrial firepower, they can only react.

What remains is a quirky WUI. Merritt Island National Wildlife Refuge, where fire managers have to cope with a fire-thirsty landscape and a fire-intolerant NASA, likes to call its setting a wildland-galactic interface. Mesa Verde has a wildland-antiquity interface. Most fire interfaces trace borders across space. Mesa Verde's intermixed landscape runs across time. This is a WUI that was laid down a millennium before. It is more protected than a legal wilderness. There is no chance to change the conditions that support the current scene. The surface woods and chaparral can't be removed, and the cultural sites can't be relocated or hardened. The effort to manage backcountry wildlands in a place dedicated to protecting cultural relics is awkward at best, and at worst ruinous. The fire program at Bandelier National Monument tried both, and in 2000 the Cerro Grande fire showed what can happen if conditions go south.

The only response—no one seriously proposes that it's a solution—is suppression. The built landscape must be shielded. But so, also, surprisingly, must the quasi-natural landscape. Researchers have argued that thinning and burning threaten the fire-sensitive pinyon-juniper woodlands, that prescribed fire will harm the trees and invite invasives, that old-growth forests deserve as much protection as cliff dwellings. The woods and the ruins are both relics. Neither has a place for open flame. Fire happens, but over such long expanses of time that it is not part of the routine creative destruction endemic to nature's economy.

What the Mesa Verde fire scene most resembles is Southern California where partisans of the indigenous vegetation and of old-growth chaparral join urban fire services in arguing for aggressive suppression. They know they cannot halt all fires, and the ones that erupt will be huge and costly. But so long as urban sprawl pushes against the land, there is little opportunity for a middle range of burning. So they argue for full-bore suppression while knowing that they will have to accept the occasional conflagration. In the South Coast that defining sprawl occurred in the postwar era. At Mesa Verde it was created in the 11th and 12th centuries. Fire officers in both locales project their anticipated fire behavior by using heavy brush models. The Mesa Verde cuesta and the Hollywood Hills make strange bedfellows.

The best preparedness strategy is those big burns. They weren't wanted, but having swept over mesa and canyon, they offer buffers against further blowups. With fuel cycles of perhaps hundreds of years, the plague of big burns has inoculated most of Mesa Verde for some time to come. The park has even incorporated them into its interpretive programs with signs and exhibits. They are, after all, the largest fraction of what its road-cruising visitors will see.

In the popular imagination the fire revolution of the 1960s was about restoring fire and living with fire. But the essence of the revolution's philosophy was to align fire management with land management. This proved tougher than anticipated because Americans frequently could not agree on what they wanted their public estate to be, or often even agree on how to talk about agreeing. But it was also based on an implicit fallacy that we could, after agreeing, craft fire to suit our ambitions. Nature, however, has its own reasons and fire its own logic. The inevitable upshot has been a series of compromises.

In Southern California it has meant substituting internal combustion for open burning, and where free-burning fire exists it comes as eruptive wildfire. In Mesa Verde it means minimum-impact suppression to prevent fires from leaping from struck snags to horizon-filling mesas, but otherwise accepting the inevitable blowups while protecting critical assets of the built landscape, although point protection is problematic when lithic points litter every square meter. It's as though a thousand years from now fire officers in the Los Angeles Basin, its human settlements abandoned for hundreds of years, had to protect the disheveled ruins of farm-terrace suburbs, celebrity pit houses, cliff-dwelling Disneylands, and car-junkyard refuse mounds against the fires that, from time to time, rush down the Transverse Range. Maybe Shakespeare was right. The past *is* prologue, even if it's in another millennium.

COLORADO

ROCKY MOUNTAIN HIGHS, AND LOWS

COLORADO WAS EXPLORED by fur trappers and the Army Corps of Topographical Engineers, but it was pioneered by prospectors. It was among the first states to replicate the California gold rush, though it showed a less cosmopolitan populace, with miners from Georgia and Missouri mingling with refugees from rushes farther West and the inevitable Old Californian who told the others how to do it. Along with farm folk marching west from Kansas, its population looked like much of the eastern United States, which is why it was admitted in 1876 as the centennial state. (The other Four Corner territories had "problem" populations—Mormons, Hispanics, Indians—that left them among the late admittees.) It retains a national mix, salted with émigrés from the East Coast and Texas, that in many ways makes it the Florida of the West.

Its mines sprouted where the ore was, in the mountains. Ranchers and farmers claimed the valleys and parks, and of course the eastern plains. Settlement was notoriously dispersed. Processing the ore gave rise to Denver, an unabashedly urban place, an echo of San Francisco. Appropriately, as entry to the arid Interior West, water was a premium. Nevada's spillover from California gave rise to the 1872 mining law, Colorado's bequeathed western water law. It was no accident that the U.S. Forest Service established its first experimental watershed at Wagon Wheel Gap in 1909.

The legacy of that sprawl continues today. Mining claims were everywhere, and homesteading for farm or ranch followed. The widely dispersed settlement, with few connective sinews, encouraged a tenacious commitment to localism. Private water works flourished, little cared for after the flush days ended. The Pike's Peak rush occurred 33 years before the Forest Reserve Act and 47 before the establishment of the U.S. Forest Service. Colorado was admitted a good 20–30 years before the advent of state-sponsored conservation as a systematic program. National forests were decreed atop that pattern. Viewed on modern road maps, the national forests bathe large fractions of the state in green. Viewed more closely, they are perforated with in-holdings whose tenure plats resemble jack-strawed lodgepole. State politics has been informed by trying to impose a general code on the resisting legacy.

Instead of disappearing into dust and playa like Nevada's, the small towns would be repurposed into tourist attractions, ski resorts, art colonies, and amenities communities generally, or reanimated by a shift in minerals. The mines transitioned from hard minerals to fossil fuels such that today Colorado has an unusually wide gamut from coal to tar sands to natural gas. Its wealth of fuels makes it 46th among the states in per capita energy costs. Drill sites litter many landscapes.

Meanwhile Denver gave it a genuinely urban concentrate that, with the spread of an aggressive car culture, began to join the chain of towns along the Front Range into a megalopolis. Western states with an equivalent urban cluster, effectively city-states, have done far better than states without one in transitioning away from an old commodities economy to a newer service one. Colorado's five largest employers are public institutions.[1]

Unsurprisingly, its fire history reflects this legacy. The wave of miners, passing over the state like a wave of human beetles, gnawed at its woods and steppes and kindled fires everywhere. By the time state-sponsored conservation arrived much of the forest had been burned. All this was quite apart from relic patterns of natural fire and indigenous burning. The argument for protection was strong. The recovered forests, however, brought new problems as monocultures often replaced former mosaics. Meanwhile, the heritage of dispersed settlement made it difficult to impose any consistent institutions other than those on the federal estate, which meant largely the national forests and the one major national park.

For long decades the recovery was adequate to hold bad fires in check. Then, as conditions changed, as forests matured and merged, as dispersed settlements thickened and joined, as climate nudged to new norms, fires revived. By the new millennium Colorado was rivaling California for houses burned and joined the top states for fire-related fatalities. Any new ideas, however, would still have to accept the stubborn heritage of local control and a relationship to the outdoors based more on recreation than on ecological integrity. None of that mattered to fire.

Colorado rediscovered it had a fire problem.

COLFIRE

The Front as Center

IN THE 19TH CENTURY, pioneered by mining, the Front Range was pitted, dammed, slashed, and burned. Then exhaustion and conservation calmed the hills, quieted the waters, and quelled the fires. The recovering forest did not burn routinely—much of it hardly burned at all for decades. By midcentury fire officers joked about an asbestos forest. Regional fires lasted just long enough to rally resources to fight them before they collapsed. The dynamics of a car-powered sprawl, quickening during the 1980s, resembled that in California, but more damage resulted from hailstorms than from fires. The chinook-propelled Ouzel fire that poured out of Rocky Mountain National Park in 1978 was deemed a monster at 1,100 acres, but good luck prevented much harm other than a loss of face and a blast of smoke. Recreation and amenities communities replaced prospecting and logging camps. Colorado seemed to have abstracted the best out of the California experience, elevated it to the high foothills of the Rockies, and escaped the backlash.[1]

Then the wheel turned. In 1989 a fire blew out of Black Tiger Gulch near Boulder and burned 2,100 acres and 44 houses ("and other structures"). The National Fire Protection Association was sufficiently shocked that it dispatched an investigation team. But Boulder seemed a separate world, as though it were a bit of California filtered and refined by long passage over the interior and the summits of the Rockies. To many the Black Tiger fire seemed less a warning than a freak, like the 1976 Big Thompson flood that followed heavy rains. It wasn't. In retrospect it was first of a mutant breed of burns. In 2000 three fires burned 22,000 acres and 69 houses. In 2002 the Hayman fire ripped a 137,760-acre tear between Colorado Springs and Denver and blew away the haze of denial. Others followed, like the plagues of Egypt. The Overland. The Picnic Rock. The 2010 Fourmile Canyon, which roared over 6,181 acres and 168 houses. The 2012 season saw breakdowns on both fighting and lighting. On June 9 a fire west of Fort Collins, the High Park, rambled over 87,284 acres and 259 houses, and killed one civilian. A week later a wildfire burst out of Waldo Canyon into Colorado Springs and destroyed 346 houses. In March 22 the Colorado State Forest Service set the prescribed Lower North Fork fire on 50 acres of Denver Water property; on the fourth day, while patrolled, it blew up, burned 4,140 acres, destroyed 23 houses, and killed three residents. It seemed Colorado could not fight bad fires and couldn't light good ones. Investigative teams were barely finished poking through the ruins when, the next summer, wildfire rambled through 14,280 acres in Black Forest, a quasi-rural exurb northeast of Colorado Springs and at 486 houses destroyed and 37 more damaged set another new standard for Colorado fire wreckage. What had seemed a California pathology had become a Colorado presence.[2]

There was no way such an outburst would not go political. The public wanted its politicians to see and be seen, and then be seen to do something. Governor John Hickenlooper flew over the Waldo Canyon fire and declared his amazement and dismay. He arranged for a committee to review the Colorado fire scene and the role of the Colorado State Forest Service (CSFS). He banned the CSFS from further prescribed burning until further notification. Legislative reforms were completed by July. Then the Black Forest fire set new records.[3]

Colorado prides itself on an environmental consciousness. Those on the Front Range live, after all, beneath one of the most majestic skylines in North America. They like being outdoors. They like open space. They want clear skies for backpacking and cycling, safe slopes for skiing, forests for greenery and hunting, summits for climbing, rivers for fly fishing and rafting, and well-behaved watersheds for domestic use and breweries. They want urban services as part of urban amenities, and the security that is a highlight of suburban life. They want nature's hazards kept in nature's mountains. They want fire cooing in their fireplaces, not roaring through their backyards.

Politically it has become a purple state—red with rural conservatives, active and retired military, and urban libertarians; blue with educated castes, ethnics, and a large public-sector workforce. Despite a string of mass shootings (that eerily parallels its serial conflagrations), it fought back efforts at gun control. It legalized recreational marijuana. So how would it define fire?

Like the nation overall, Colorado's private lands are overwhelmingly in its eastern half, and its public lands in the west; along the Front Range they mix. The federal government oversees about 36 percent of the state's land mass; the State of Colorado, less than one-half of 1 percent (three million acres). Those federal lands, however, contain 68 percent of the state's forests. Wildland fire responsibility resides with the local governing authority, whether land agency, fire district, or city. For places outside those domains the county sheriff has jurisdiction. It's not an unusual arrangement, and before it was stressed beyond fatigue points, a reasonably workable one.

But fire was never a high priority. When the state decided in 1911 it needed some agency accountable, it established an Office of the State Forester under the College of Forestry at Colorado State University. In 1933 it consolidated the state forester and the State Land Board, then reversed itself four years later, then reversed again in 1945, and back once more in 1955 when the office became known as the Colorado State Forest Service, jointly directed by the vice provost for Agriculture and University Outreach and the dean of the Warner College of Natural Resources. In 2001 its charges expanded to include service to the Division of Forestry in the Colorado Department of Natural Resources, created two years earlier.[4]

Its charges involved more than fire, but it was designated as the State's lead agency during suppression. It also maintained a self-funded fire equipment shop and trained in prescribed fire. In practice, local jurisdictions fought the fire, and if they requested help (through the sheriff) the CSFS would respond, mostly by further delegating to interagency incident management teams. It had neither the capacity nor the desire to command fires for any length of time. The reality was that jurisdictional responsibility was as pocked with land tenure as with in-holdings. Most wildland fires occurred on federal land or on lands with cooperative agreements with federal agencies. The escalation in fires and burned area (and housing losses) that has occurred over the past 30 years has been primarily on state and private lands. The federal lands were among the least funded among the federal agencies, but they had access to national resources, and agreements with the state and counties extended that same access to them. The CSFS was a bureaucratic switching yard.

Then everything bulked up. Small towns became large, large towns swelled into metropoli, fire's one-day wonders morphed into campaigns, and what had been annoyances became catastrophes. The jump occurred around 2000. The Roosevelt National Forest, roughly from Boulder to the Wyoming border, nicely illustrates the phase change. From 1885 to 1999 some 398 fires burned an average of 14 acres. From 2000 to 2015 810 fires burned an average of 110 acres, or if one subtracts the High Park fire, an average of 57 acres. The fire scene had increased twofold in starts and fivefold in area. People started about twice as many fires as lightning and burned twice the acres (this does not count 200–300 abandoned campfires).

But the real story was happening on state and private lands. The 1960s averaged 457 fires a year that burned 18 acres each. The 1980s more than doubled the number of starts (1,286) but held the average acreage to 18. The 1990s increased ignitions by 50 percent (1,806) and actually reduced the average burn by a third (12 acres). Then the lid blew off. The new millennium saw an average of 2,973 fires for an average burn of 39 acres. Incinerated houses followed the same curve. The jet assist came with the big burns, which were also the ones that took out the most houses. Colorado's 30 largest fires on record have occurred since 1996, 77 percent since 2002.[5]

The new wave of fires scared Utah. They scarred Colorado. The old order failed its stress tests. To public and political eyes the CSFS could neither fight fires nor light them, immobilized by the trauma of the Lower North Fork fiasco. That left an eager urban fire service dominant, and they and politicians defined the issue as one of public safety. They wanted a single agency to coordinate all-hazard emergencies and to be a point of accountability. They wanted to know who would respond and who might be blamed.[6]

The upshot was to transfer wildfire responsibility from CSFS to the Colorado Department of Public Safety and to ban, until further notice, any state-sponsored prescribed fires. The CSFS shrank into an outreach and extension program. What fire staff and equipment it had went to the new Division of Fire Prevention and Control (DFPC). The DFPC had statewide responsibilities, though jurisdictions at a fine grain remained murky. It expanded from 25 to 110 personnel, from four to nine engines, kept two single-engine air tankers, acquired two multimission aircraft, and contracted for three helicopters. In 2013 it received authorization for a Center of Excellence for Advanced Technology Aerial Firefighting located in Rifle. But it had no enforcement powers. Colorado wanted the state to supplement not supplant local control.[7]

It was not an unreasonable argument. It's a foundational premise that fire management must bond with land management, and if the primary land use is urban or exurban sprawl, then an urban fire model makes sense. If California defined the fire problem before anyone else, it had also longer than anyone else worked out solutions. It opted for an urban fire service, even if it operated in the woods. The California Department of Forestry morphed into CalFire. A casual observer might conclude that Colorado was moving, with its own quirks and twitches, along the same path, that he was watching ColFire in embryo.

It's an attractive choice, politically. It connects visibly with voters. The state is seen to be taking active measures. This is not, however, the vision of the National Cohesive Strategy, of fire-adapted communities living next to or amid fire-resilient wildlands. It's a vision of fire-protected enclaves like embattled urban fortresses in a hostile countryside. It's an attempt to solve an immediate crisis—was defined as a crisis in public safety—without addressing the fundamentals underwriting those explosions. It relied on the Front Range's traditional understanding that fires would be brief and episodic. Damages might be graphic, but they were limited compared to the scale of the urban scene. Fires ran bigger and longer but were still relatively ephemeral problems compared to enduring conundrums like transportation. Forest health mattered less than Medicare and health insurance. Colorado DFPC was an all-hazard emergency service that suited the state's preference for local control and dispersed political patronage. Clearly, wildfires along the Front Range were the casus belli, but the DFPC had statewide responsibility.

The less visible issues are three.

One, the strategy divorces firefighting from land management. It relies on fire suppression to contain fires without the urban zoning, building, and fire codes that had abolished fire from cityscapes and have left most urban fire services with far more medical responses than fire calls. That choice for codes belongs with local jurisdictions, a few of which (like Jefferson County) have adopted rules for defensible space in new construction. Instead, DFPC is charged with containing fires without the capacity to control the combustible cityscapes or WUIs that carry those fires. Coloradans wanted incentives, not enforcement. They wanted carrots, not sticks. The question was whether fire, which ignores carrots and feeds on sticks, would agree.

Two, the Colorado strategy adopts, if sotto voce, a Southern California strategy without equivalent Southern California muscle. Four of the five largest fire departments in the United States share the South Coast. Colorado's Front Range, however, has the spread of Southern California without the massive resources to respond to multiple structure ignitions or fires that can burn for days in and out of the urban fringe. It would have to rely on very rapid detection and attack rather than overwhelming force on fires that broke free. It would substitute technological wizardry for massive response. It would appeal to federal agencies for backup.

And three, it has disabled the state's ability to manage fire, including prescribed fire, across landscapes rife with dispersed settlement. The ban against prescribed fire was lifted for pile burning in late 2014, and for some broadcast burning in 2016. Without the ability to dispose of fuels, however, mechanical treatments for defensible space will choke on their own woody effluent. Mitigation remained with the CSFS, which functioned like an extension service, offering advice, assisting with plans, continuing to funnel funds (mostly federal), and generally facilitating projects. Beyond home ignition zones, there was the question of fuel treatments and ecological health (for example, for state parks and wildlife sites) at the scale needed. Outreach mattered, and it had its successes, but it was not at all clear that it could move faster than fire. If not with tamed fire, the land would burn with feral fire.[8]

In sum, in the absence of compulsory codes, or commanding resources, or the capacity to manage landscapes, Colorado hopes to substitute rapid detection and response to hold outbreaks while small. Instead of the Cal-Fire or the Los Angeles County Fire Department tendency toward overwhelming force—dispatching strike teams of engines, bulldozers, and air tankers at first alarm when conditions warranted—it would rely on technology to find and hold fires early. The Center of Excellence continues to explore night aerial firefighting, drones, satellite messenger evaluation and air-to-ground data links. It promoted Colorado-based mapping technologies and a Colorado-specific fire decision support system. The goal was to substitute nimble aerial technologies for on-ground heft. The DFPC bet it could catch fires before they became problems. It's urban-style fire planning projected onto exurban landscapes.

The agency has little choice. There is scant social or political license to explore more robust models. The urban Front Range has zero tolerance for fire or smoke. As Forest Service fire management officer Bryan Karchut put it, the Front Range has evolved from a 10 a.m. policy to a 10 p.m. policy in which every fire would be caught before the end of the day.[9]

Justice Louis Brandeis famously declared that America's states were the experimental labs of democratic government. Colorado's experiment is to adopt an urban all-hazard model without urban-style preventative measures and to hope that lean and mean suppression will substitute for large and varied landscapes and a political context that wants maximum results from minimum investments.

It's all a work in progress—the ink is hardly dry on the drafts, everything is an interim measure that will have to work toward some equilibrium with Colorado conditions. The reforms clarify the fire scene on the Front Range, though not in a way that may be sustainable. They leave fire codes with the locals and fire management with the feds. Waldo Canyon rebuilt to code. Black Forest did not. The Colorado Division of Fire Prevention and Control has to broker between the two.

The three-pronged strategy of the National Cohesive Strategy—fire-resilient landscapes, fire-adapted communities, better capacity—accepted such a world and accepted that fires would happen (and would need to happen). The Colorado strategy also accepted that world but believed it could keep fires out. It did not write fire-adapted communities into the DFPC's charter. It said nothing about fire-resilient landscapes. Colorado beefed up suppression capabilities because it just wanted those fires stopped.

Colorado's landscapes, like its politics, are split, and it is unlikely that one side or the other will prevail soon. Fire-prone houses will coexist with fire-prone woods, and conservatives with progressives, not mixed but layered like oil and water. Like managed wildfire in which high-value assets are protected and a fire loose-herded elsewhere, the Front Range would get full suppression and the federal backcountry could get some kind of fire management.

That strategy assumes clear boundaries, but Colorado's legacy of land tenure has bequeathed a jumble of jurisdictions with little chance to impose or inspire some governing authority over them all, apart from crises of fire suppression. The escalation of fires and damages occurs mostly on private lands. The resources to control fires reside mostly on public lands. The choices of the citizenry tend to leave local control without local capacity. The situation is metastable, and will remain so as long as the wild and the urban mingle promiscuously, and residents want to pretend an oft-explosive nature is a recreational park.

FIREBUGS

THE NEW MILLENNIUM ushered in, along with fires, a wave of wildland insect infestations. A few were exotic, such as emerald ash borer and sudden oak death fungus. Most were indigenes such as spruce budworms and beetles that, like locusts, found favorable circumstances that allowed them to swarm over and devour vast tracts of land. The mountain pine beetle (MPB) has been most the notorious of that mob, as its latest infestation blew out of the interior of British Columbia and slow burned, in a 10-year front, throughout the American West. The beetles spread like a Biblical plague until the lands they altered rivaled the size of the Laurentide Ice Sheet. Then the front passed, its fuels exhausted, a few beetle embers aglow.

The beetles resembled fires in some ways. They burned patchily. They fed on the lush, often monocultural fuels left by settlement and that had excluded fire, and they were leveraged by climate change. In some places they picked off the equivalent of a snag or two, or torched a grove, while in others they took out whole stands. They were, for pines, what Spanish flu was to humans a century earlier. One epidemic encouraged another, as spruce beetles came on the heels of pine beetles. For the new era these outbreaks were the equivalent of megafires. Among the worst hit regions was the Arapaho-Roosevelt National Forest along the northern Front Range of Colorado.

Here, it seemed, was another anthropogenic insult to the land. The same lineup of causes that fed fires fed beetles. Worse, the beetles were converting live fuels to dead. Their expansive effluent would evidently fuel a new era of ravenous fires. The Four Horsemen of the Anthropocene were colluding among themselves.

But how do beetles and burns interact?

It's a sticky question; and as with fire, we know mostly about mountain pine beetle because it concerned foresters, who saw infested woods as lost timber and combustibles for conflagrations. Fire and beetles were thus studied for similar reasons: they were a threat that needed to be controlled. Ideally both fires and beetles would be attacked and contained as quickly as possible; forestry agencies had a similar policy for both, one that failed. Moreover, the prevailing view was that fire and beetles encouraged one another. Burns created conditions favorable for beetles, and beetles, by transforming wet live fuels into dry dead ones, stoked fires.[1]

Anyone who walked the woods could see this. Particularly with lodgepole pine, beetle kills left whole necropoli of dead pine, both standing and fallen. The severity index rivaled the worst blowups; on the Arapaho and Roosevelt National Forests between 50 and 90 percent died. The fuel loads ran into the tens of tons per acre. Grasses, shrubs, conifer reproduction overgrew the logs—"lodgepole chaparral," as fire behavior analysts mutter. Those killing fields are slash piles ready for the flame. More fuel, more fire.

More nuanced research presents a different perspective. The flammability of a forest following beetles is a rolling target. When the trees still hold dead red needles, the crown fire hazard increases, as whole hillsides seem to have rusted in the rain. When, after a year or two, those needles fall, the hazard plummets. When, many years later, the dead trees themselves fall, the fire hazard again increases but in a different way, as boles interlace with grass, shrubs, and reproduction, then mature into yet other fuel arrays. When drained by drought, those heavy fuels can burn stubbornly. Their jackstrawed piles make travel onerous and firelines hopeless, and the danger of still-falling snags serious. The character of fire changes. Landfire maps become useless. The past appears as a palimpsest, partly erased and written over. Most of the scene is not explosive, however, and not all burns are severe. As with fire the beetles can only tweak what is already there.

The 2012 High Park fire ripped across the Roosevelt National Forest through ranges laden with beetle-killed pine. Postfire surveys found that in lodgepole pine forests, the proportion of the landscape burned in "each severity class was similar for healthy trees (dense veg) and recently killed trees (early stage MPB)," while burn severity in "late stage MPB was slightly reduced." The same proved true for ponderosa pine and mixed conifer forests. The big difference was between the fuel characters of the foundational woods. Lodgepole burned more severely than ponderosa, which it does with or without beetle infestations. What mattered most was "forest density."[2]

In other words the changes in forest structure wrought by settlement outweighed the changes from recent beetle outbreaks. Even the severity of the beetle outbreaks might be attributed to the upheavals set in motion by contact and its ecological aftershocks. Another study across western North America found that "insects generally reduce the severity of subsequent wildfires." They might even "buffer rather than exacerbate fire regime changes expected due to land use and climate change." What looked bad didn't necessarily burn bad.[3]

If this seems profoundly counterintuitive, it may be because we are conditioned to thinking about fire only in terms of fuel and flame. We aren't really thinking ecologically. Free-burning flame and slow-metabolizing beetles are two expressions of the same chemistry. They compete for the same fodder. They combust it, at the molecular level, by the same reactions. Unlike goats the beetles don't chew the same particles that fire would burn, but they can kill the tree that carries those needles and twigs. In a biologically based theory of fire on Earth, we could find ways to record their relative contributions and interplay. We would recognize the ways in which beetles are vectors for catalyzing complex ecosystems within trees. We would see the futility of attempts at wholesale suppression through mechanical and chemical means (sound familiar?). Of course we can translate the beetle-inspired changes into fuels, but that seems to miss the more interesting dynamics grounded in biology.

Less abstractly, we may be seeing with beetles what we saw a century before with livestock. In the arid West settlement unleashed a plague of cattle, horses, and sheep. Initial contact often sparked fires to help green up the range. Then the fauna ate fire out of the landscape and fires became rarer. All this changed the character of the surface vegetation, which morphed from grasses and forbs into the shrubs and conifer reproduction that has helped flame leap from the surface to the canopy and has resulted in an epidemic of high-severity fires. Might something similar be happening with insects?

Fire, fuel, and fauna, even small or microscopic fauna, make a three-body problem for which there is no exact solution. Fauna move, and people can move both fire and fauna. The animals can move or be moved by humans; they respond to social conditions as well as to climate and terrain. So, too, fauna with or without people respond to dynamics other than fire and fuel. They interact. Fire can stimulate as well as consume fodder, and fauna can not only eat fire-favored particles but help create the conditions that feed fire. What emerges is a complex choreography that makes sense ecologically but doesn't always make sense if reduced to the physics of fuel loads.

Unlike cheatgrass mountain pine beetle is not an exotic, which can resemble a software virus busy rewriting the operating system for sage steppes. Unlike cattle it is not something that gobbled and trampled whole ecosystems, or that can be rounded up and herded away. The beetles are natives; they exist in a mutual accommodation with the woods, though like juniper they now find themselves moving into new sites as opportunities warrant. Except on small scales (say, trees used as landscaping) direct control doesn't work.

In 2016 the beetle epidemic was fast ebbing. The vulnerable woods had been taken. The legacy lived on, notably in the modified landscapes left behind. Colorado had two large fires that showed what that could mean for fire management. The Hayden Pass fire burned through red-needle, beetle-killed spruce—a vigorous but clean fire. The Beaver Creek fire burned through old beetle-killed lodgepole, now lumbering the landscape like the aftermath of a blowdown, leaving a coarse-boled lattice through which new conifers, grasses, forbs, and brush intertwined. It would take many hard, expensive, smoky months to contain the fire's perimeter, with falling trees more a hazard than flames, and new restrictions (like the threatened sage grouse) that denied the traditional Plan B, to fall back to sage steppes and burn out. Or to take a longer view, since at least the mid-19th century, the Front Range has never had the chance to learn how to cope with a recurring kind of fire in a way that allowed incremental improvement.

Or more broadly still, the ecology of how fire, fuel, and fauna interact is always indeterminate, a dynamic of jostling. We have to learn to live with beetles as with fire, and more awkwardly we have to live with them both and their interplays at the same time.

Unsettled, too, is the legacy of their interaction. Beetles and fire, slow burn and fast, are an old tag team. But in the new millennium they have been operating on scales previously unknown. (It's like watching old comic book characters [think Archie and Jughead] morph into today's Hollywood heavies [Incredible Hulk and Ironman].) The beetle-killed forests burn in ways that are effective in reducing the dead boles to ash, but over long periods and sometimes a long-lingering venting of smoke. They are hard to contain in the backcountry. Combined with other global disorders, they may spark new biotas. Or not. We're in an experiment in which two natural agents are interacting on a scale not previously known in circumstances not understood. It may result in novel arrangements of the old ways. Or it may create new landscapes with fire properties quite different from what we are accustomed to. The fire community may find that the old ways don't work as they used to because their context has changed, and will likely keep changing.

We are still in early days of integrated landscape management. We have a need for more sophisticated ecological responses, which means not just imagining landscapes as firesheds but as fire habitats. Landscapes are more than stockpiles of fuels; they are places where fire, as a product of the living world, interacts with other processes and creatures. In this world action doesn't lead to reaction, but to tangles of interactions. We need to think of biotechnologies, techniques beyond snuffing flame with water and chemicals. There is a place for prescribed grazing. Might there sometime be a place for prescribed infestations? Beetles are species-specific. Can we imagine kinds of biotic inoculation that could alter fire behavior without resorting to mechanical treatments and herbicides? At present it's only a thought, or to skeptics maybe a hallucination, and to critics probably a Frankensteinian horror.

But the day is coming when we need to think seriously about fire's biology, not just the physical metrics of its spread and severity. Fire is a product of the living world—and this remains its most fundamental habitat. We manage fire by managing that habitat. The habitat is more than blocks of hydrocarbons waiting to be kindled or quenched. The beetles know this, and know how to live with fire as both rival and stimulus. Maybe that is what their latest outbreak is telling us. Call it the fire sermon of the beetles.

THEN AND NOW, NOW AND TO COME

IN THE SPRING of 1983 I was restless. I had spent every summer of my adult life with the North Rim Longshots. But that had ended in 1981. The summer of 1982 I lived in Iowa City, far from anything familiar, writing *Introduction to Wildland Fire*. That winter—the austral summer—I went to Antarctica, which served as a surrogate of sorts. Now I was looking at another out-of-sorts summer when I heard from a ranger I had known at the Rim. Steve Holder had gone to Rocky Mountain and was there when the Ouzel fire blew up. The park needed to do something about fire. He suggested I come and write a plan. The regional office would fund it. I said yes.

There was far more interest outside the park than within it. Dave Butts, then NPS director for the Branch of Fire Management, had written a master's thesis on fire at Rocky, had a cabin outside it, and wanted the park to recover from the embarrassment of Ouzel. Jim Olson at the regional office thought Rocky needed someone to manage fire and vegetation. Rocky's neighbors—private landowners as well as the Roosevelt National Forest and various county sheriffs—wanted a credible fire program. The park thought fires so rare and Ouzel such a fluke that it was pointless to waste money and mental bandwidth planning for something that would never happen again. The park had a fire plan that had failed. It had no fire officer. The park's ongoing crisis was its crushing visitation. Its principle resource management concern was its elk population.

I spent that first summer studying the park and what fire literature existed. Because Colorado State University was an hour away, there was a fair amount of literature on fire ecology and history, or so it seemed at the time. There was enough to frame the conditions for fire management. I ended the summer with a 20-page summary of observations and ideas. I thought that with another summer I could write a formal plan according to NPS-18. My fundamental conclusion was that Rocky did not have a wildfire problem, but it did have a fire management problem.

I returned in 1984. I wrote the plan on my personal PC, a 286 computer with an IBM green screen monitor, and printed it on a dot-matrix machine. The accompanying maps were hand drawn and colored with pencils. (These were days before the digital revolution. A man showed up one day at our Park Service housing to ask if we wanted a TV connection. We said, No. Why? We had no TV. He asked if he might use our phone to make a local call. Sorry, we replied. Why? We had no phone.) Since the park had so little capacity, I spent some time with the Arapaho and Roosevelt National Forests drafting a cooperative agreement that effectively turned over fire response to the forest, under a suite of guidelines to keep any actions within the mandate of the Park Service.

The park refused to outsource its fire program—no surprise there. My recommendation for a dedicated fire management officer, even fused with Jim Olson's interest in a dual appointment for fire and vegetation management, sank in the sands. No fires repeated Ouzel. No prescribed fires were lit. (The wildlife biologist was horrified at the idea because "the elk will be all over" those burns.) Nationally, the years 1982–84 were among the lowest for burned area on record. Rocky Mountain National Park was an asbestos park with an asbestos forest, and the sense was that the park might as well worry about a potential meteoric impact as about a big fire. Conflagration is a classic low-odds–high-impact event.

When, 30 years later, researchers with the Public Lands Program at Colorado State University inquired into Rocky's fire program, no one could even find a copy of the 1984 plan I had submitted. It had shown less staying power than Ouzel's smoke, which had at least elicited a citation from Boulder County for violation of air quality. I spent the following summer in Yellowstone, a place with more fire but equally uninterested in modernizing its fire program.

In August 2016 I revisited Rocky. Much had changed, the basics hadn't. Visitation, wildlife, and scenery remained the dominant concerns. Proximity to the Front Range (and the first mountain park for flatlanders from the Midwest), Rocky Mountain was now the third most heavily visited park in the nation. The wild and the urban were like charged anode and cathode, ready to arc. Nearly all of the park was legal wilderness; a third of its outside border was private, and rapidly urbanizing. Estes Park and exurbs along Rocky's Front Range were unrecognizable. But so, more astonishingly, was its fire program.[1]

The Yellowstone fires of 1988 had poured money into fire management in the NPS. The 1994 season widened interest, and Interior became a major force for reform. The buildup began, even at Rocky. By 1996 it had a dedicated fire management officer and assistant and two fire seasonals. The program tackled some fuels issues along the developed areas. When a prescribed fire annoyed Estes Park with smoke, they were told to shut it down. They shut down the program. Then the floodgates opened with the 2000 National Fire Plan (NFP). The NFP lavished funds particularly for hazard fuels reduction around wildland-urban interfaces. Rocky may have had few fires, but it had a serious WUI. It could claim a lot of attention and dollars.

It now has a full staff—fire management officer, fire ecologist, fuels specialist. For fire suppression it has two six-month seasonals, and one permanent subject to furlough. For fuels it has two permanents and 10 seasonals, and the intermittent use of a hotshot crew. Much of its fuels work, however, it contracts out. It has a complement of vehicles, including an engine. It has two dedicated buildings, one a fire cache, the other equipped for 10 offices. It hosts the Alpine Interagency Hotshot Crew. Fire science has matched the buildup, with more detailed atlases of fire history and understanding of ecology, especially as mountain pine beetles have slow burned over the park and neighboring forests. In heavily infested areas as much as 70 percent of trees died. In 2012 the Fern Lake fire repeated the Ouzel scenario and ran wildly before 70-mile-per-hour chinook winds and this time burned a house and forced Estes Park to evacuate. Another fire that year, Woodland Heights, wholly within the town, burned 20 acres and 22 homes. Wildland fire had become serious business. It had the attention of park authorities.

I felt like Rip Van Winkle, utterly disoriented from the Rocky I had known 32 years earlier. In my absence a revolution had taken place.

Yet the basics endured. There was a fire source, wildlands, and a fire sink, towns and exurbs. It's hard to imagine that fires in the montane forests such as Estes Park had not been routinely burned for wildlife in presettlement times, and that some of those fires would not have run into surrounding hills. Settlement boosted that burning and carried it into the woods. By the time the park was established, much of it had burned, repeatedly. Frederic Clements researched his famous "fire history of lodgepole burn forests" from Estes Park. But over the last century one large fire—large being several hundred acres—had broken out on rough average every 15 years. What made those burns dangerous was the late-fall chinooks that could break loose even modest fires into avalanches of flame. Those winds joined source to sink.

The source was changing, as climate, beetles, and the legacy of unburned woods bounced off each other. In 1984 only 1 percent of Rocky was legal wilderness; in 2009 95 percent became wilderness, leaving the source to develop according to mostly natural forces over which managers had little say. By almost any measure the scene was worsening toward explosive burns. So, too, the sink was deepening. More dispersed exurbs, more broadened suburbs, more expensive houses, most outside effective fire codes. Eventually source and sink will fuse. As in earthquake prediction, the identification of place matters more than that of time. The source is intermittent, the sink constant. The house odds favor the sink over the source. In a rational world that's where one would concentrate efforts.

Rocky is Rocky. It's a national park precisely because it is unlike other landscapes. So how might its fire scene reflect national concerns?

Rocky shows how fire has come to claim national attention, even at places not routinely threatened. Residents may be in denial over their particular dwelling, but not about the country's quickening pace of fires. Rocky demonstrates, too, the astonishing improvements in equipment, technology, science, staffing, and training that have occurred over the past two decades, particularly since the National Fire Plan. Because Rocky began with nothing, the fire program's buildup is undeniable and easily measured. That buildup came from pushes and pulls outside the park itself. The park's fuels work is serious; it has even managed to maneuver into fringes of the wilderness by committing to use only manual tools.[2]

Yet it's not enough. Its master plan for fuels estimates that at current rates the park will require 65 years to complete a full sweep of identified areas, and that only pertains to the park itself, while the hazards lie mostly beyond its borders. In good American style the present arrangement is costly enough to be noticeable and irritating but not enough to do the job. And even if the park could complete its modification of the source, the sink is unlikely to match it. All this sounds a lot like the national scene. The National Cohesive Strategy tries to balance fire-adapted communities with fire-resilient landscapes, but at Rocky, as along the Front Range generally, there is little interest in fire-adapted communities and scant opportunity for fire-resilient landscapes.

The only way out is collaboration, not just for money but for shared resources, risks, and responsibilities. Elsewhere, alliances exist among NGOs, chambers of commerce, wildlife organizations, municipal watersheds, federal land agencies, state fire organizations, homeowner associations, city governments, fire protection districts. There are plenty of places nationally that are moving in this direction, whether or not they are moving fast enough or broadly enough.

It's what Rocky Mountain National Park must do. It can coax some fire on its west side, where the stony summits of the Rockies etch a broad fuelbreak. It can manage some fires in the northeast, as it did with the Cow Creek fire in 2010, though the long-lingering plume of that burn—so visible along the Front Range—made residents uneasy. But this is not operating at scale, even at the leisurely big-burn pace historically true for Rocky, and it is not protecting communities. It isn't burning the mountain parklands that historically must have burned regularly.

The fire staff understands very well the dilemmas it faces. It frets over matching means and ends. It struggles to get good fire back into a place dedicated to natural processes. It worries about coping with bad fire, or in the words of fire officer Mike Lewelling, about an Estes Park down "the gunbarrel of 70 mph winds." With both good fire and bad relatively scarce events it's hard to plan, hard to program, hard to manage. Instead, the staff must play the odds and hope that what it has done will align with what it needs when the crisis hits.

I won't be around in another 32 years. But I wonder what fire at Rocky will be like in 2048. And where I should place my money.

FATAL FIRES, HIDDEN HISTORIES

COLORADO IS NOT what comes to mind when most observers think of wildland fire fatalities by burnovers. The giant is California—a universe unto itself. Then comes Idaho, but its figures are distorted by the Big Blowup; the era of record begins in 1937. Then comes a cluster of states that includes Montana, Arizona, and Colorado. What makes Colorado's ranking particularly intriguing is that the two major events, both involving hotshot crews, occurred in scrub oak, a fuel not normally considered volatile. But at Battlement Creek in 1976 and again at South Canyon in 1994, frost-killed Gambel oak powered an explosive run that trapped crews.

Each resulted in reforms. The Battlement Creek burnover led to an emphasis on personal protective equipment, especially fire shelters, which became mandatory. The South Canyon fire set into motion a series of corrections that made firefighter safety a primary goal of wildland fire management. It changed how fires would be fought. Then, more slowly, nudging the system the way a locomotive might crawl around a tight curve, it changed what fires would be fought.

SOUTH CANYON: A PROSE ODE

"No place, not even a wild place, is a place until it has had that human attention that at its highest reach we call poetry."
—WALLACE STEGNER[1]

For a fire historian the treks can take the form of a pilgrimage. I have been to the Big Blowup and Blackwater, stood before the memorial at Bass River, looked along the ridge at Rattlesnake, into the ravines at Inaja and Honda Canyon, at the hillsides of Loop, Battlement Creek, Yarnell Hill, Mann Gulch, and now South Canyon. It's different being there. The slopes seem steeper, the scenes more compact, the setting more intense. Most of the burns seem surprisingly small. Most sites today are very quiet.

When I was there one bright August noon, the memorial plaque overlooking the South Canyon blowup was empty of other people. A light breeze skipped through grass. The sun passed over white-trunk juniper, dull as bone. A fly. A bee. Lizards. A landscape haiku: stone, a burned stump, weeds. There is no shade, there is no relief. It's a place at which to contemplate but not at which to be comfortable.

But what exactly to contemplate? The interpretive signs that line the trail speak to the job of firefighting. They speak only of doing, not thinking. Yet everyone dies. People die at their jobs. They die at home. They die in mines. They die in vehicle accidents. They die in floods, tornadoes, hurricanes. The wildland fire community has never been known for its meditative powers. It's a community that thinks with its hands. In 1994, when the South Canyon fire blew over a mixed crew on the slopes, 20 other firefighters also died that year. Why memorialize these 14? Why this way?

One answer of course is that this disaster took the lives of four women, which made it a novelty. Another answer is that those earlier crew disasters did get attention, though nothing like that granted to the South Canyon dead. Twenty-three years after the Big Blowup, the unclaimed fallen got a dedicated burial circle and commemorative stone at Woodlawn Cemetery in St. Maries, Idaho. After Blackwater, the U.S. Forest Service and American Forestry Association created a medal for forest firefighters killed on the job. At Mann Gulch concrete crosses were erected.

Generally, the dead were viewed by whatever lens caught the prevailing sentiments of the day. The fallen of 1910 were workers, hired by the government, who deserved a decent burial. The fallen at Blackwater were CCC boys, whose death reminded the country that they were engaged in a great experiment to rehabilitate a society ravaged by the Depression and a land wrecked by heedless settlement; regrettable, accidents happen, sometimes "without fault or failure." The fallen at Mann Gulch got a spread in *Life* magazine. Memorializing the Rattlesnake dead was tricky because they were temporary hires from the New Tribes Mission, a Bible college, and the fire had been set by a man looking to be hired as a cook for the suppression effort. The Inaja dead—most of them an inmate crew—were memorialized by Chief Forester Richard McArdle as having died "in the defense of the free world." The Loop fire fallen, the El Cariso hotshots, had worn berets and could be seen within the flawed buildup of the Vietnam war. The Battlement Creek dead were accident victims for whom better equipment might have prevented their loss, as though the wildland fire community needed the equivalent of airbags.[2]

In 1992 Norman Maclean changed how we would remember dead fire folk. He could appeal to deeper, even existential concerns, and allude to cultural talismans well outside the scope of the fire community. He changed the terms of how we understand firefights when something goes lethally wrong. *Young Men and Fire*—published two years before South Canyon—demanded that we recognize each of the fallen as an individual. It saw the fatal moment not as an accident, in which unlikely events randomly colluded, but as a tragedy, a clash of wills, as a crew attempting to control a fire met a fire that would not be controlled. What had been labeled "disaster" fires now became "tragedy" fires. In this way his meditation spoke to universal truths of the human condition.

Every fatal burnover since then has been interpreted through the prism of his text. All the old fires have been revisited and rememorialized in the vogue of South Canyon. On the occasion of its centennial even the Big Blowup had a new commemorative stone installed amid the burial circle. In fire, as in geology, it would seem that the present is indeed the key to the past. The Big Blowup was remembered on the job by the pulaski tool, a means to do the work more efficiently. South Canyon would be remembered with a change in practice, coded into norms and mores.

Maybe it's time to move on. If *Young Men and Fire* added a new narrative to American fire, it still told the story, once again, of a firefight, though one gone bad. It did not ask whether or not that firefight should have occurred. The jumpers at Mann Gulch stood for all of us. We all die. Mann Gulch itself offered literary possibilities that made a blowup fire that crashed through the crew an attractive symbol for the fate that awaits each of us. The traditional response to disaster fires has been to double down, finish the job, man up, make sure the fallen had not died in vain. The response to tragedy fires is to not forget.

The literature on fire seems so haunted by Maclean's masterpiece that it appears unable to move beyond recycling the story, often not even speaking to the Mann Gulch fire so much as to Maclean's telling of it. It's the literary equivalent of doubling down, of insisting that the dead be not forgotten by doing better, telling the story with new information and fresher insights, like finding a better combi tool or a leadership rule of thumb. For a community committed to doing, however, one can wonder not only if retelling can replace doing, but if the theme is an adequate narrative for what the fire community must now do.

Before, responses were always about doing the job better and safer, not whether or not it should be done at all. Blackwater led to reforms in the CCC and better crew organization. Inaja to the 10 Standard Fire Fighting Orders. Battlement Creek to personal protective gear and fire shelters. South Canyon to making firefighter safety an informing principle in plans. Not until the Yarnell Hill fire was the issue raised formally—and then by Arizona OSHA—that some fire settings are so toxic that there is no justification for having anyone on them at all.

The trail of plaques at South Canyon continues the older logic. It would ask too much that those reeling and mourning should dispense with this long tradition or with the closure and maybe catharsis it might bring or to deny the impulse to *do* that lies at the heart of the culture. So it would seem disrespectful to the point of moral callousness to suggest that a fire whose location was first misplaced, a fire left for three days before any action was taken, a fire that did not threaten vital resources or communities might not have needed an undercut line through dense, frost-killed Gambel oak.

Maybe the whole ethos is misguided. We made a wrong call back in 1910, and cannot return to those times to correct it. Instead, we renew those commitments so as to memorialize those who died trying to make it happen. Then, when the story goes south again, we search for catharsis. But if we need catharsis, maybe we're doing the wrong job, refurbishing the wrong story. Maybe it's time to pull back from the flaming front. Maybe we need to go indirect on the narrative.

That won't happen anytime soon. Instead we will honor the dead by learning from the compounded errors and ensuring that others don't die as they did. A narrative that speaks to fire as a coming of age story, that replaces fire fighting with fire managing, that can remember the fallen as men and women who did not die in a way of their choosing but did die doing what they chose, will have to wait for another time.

Not all places are glamorous, not all events sublime, not all lives heroic. We all die, or as Ernest Hemingway, a contemporary of Norman Maclean, put it, every true story ends in death. Some deaths and some stories seem particularly compelling.

Death by fire, in full-throated roar in wildlands, is one of those that can grip us by the lapels, get in our face, and shake us out of our routines. It appears to burn away the veil of normalcy, exposing a tear in the space-time continuum by which we understand the quotidian world. The usual tapestry of tropes, anecdotes, clichés, habits, and expectations is ripped apart. We need explanation. We need catharsis. We need closure. We need the rend mended.

That's what interpretive trails like the one at South Canyon do. By speaking to doing the job better, and by remembering the fallen with photos and bios etched into plaques, they close the hole. They suggest that the normal can return, just better equipped, better trained, better led, making sure that that vital weather forecast doesn't get mislaid or those spot fires in the ravine aren't overlooked, that the fire next time will not kill those sent to cope with it. The message is that we might be among the saved. The fate that awaits us all will not be in the flames of perdition.

BELOW THE FLAMING FRONT

Even celebrity events can have their hidden histories. There is no one way to understand the past. We know it through context, and as that setting shifts, so does our appreciation for what something from the past means. Mostly we rely on interpretive templates to contain free-burning intuition. With fatality fires, these are typically technical that focus on how fires happened (and were fought) or sentimental that say why we should care. But since these events are fires, they also have a context in the larger fire history of Earth. That holds for the two fatality fires for which contemporary Colorado is known.

The South Canyon fire burned outside Glenwood Springs. But another fire, origin unknown, has been continually burning since 1910 in the coal seams outside town. Mostly it's a case of out of sight, out of mind. It's known at the surface through subsidence patches, venting gas and ash, condensates, and red oxidized shales. If the minieruptions from South Cañon Number 1 become troublesome, crews fill in vents with dirt. In June 2002, however, when the West was hunkered down in drought and big burns, the surface venting kindled a fire that roared over 12,229 acres, incinerated 29 houses in West Glenwood Springs, and cost $2.5 million to finally suppress. Then August thunderstorms triggered mudslides. Burned slopes sloughed off soil and detritus that rushed into Mitchell Creek and blocked I-70.[3]

The Battlement Creek fire crossed another narrative of fossil fuels. The Mesa Verde formation is dense with gas, sandwiched between sandstones and shales. If it can be fracked, the gas will flow, but old-style fracturing was slow and costly, so the scheme bubbled up to use 40-kiloton nukes. Project Mandrel Rulison, one of three test sites, combined underground nuclear tests (Operation Mandrel) with Operation Plowshare, a broad enterprise to turn atomic energy to peaceful uses, and commercial uses through the Austral Oil Company. (The idea probably belongs with other delusional proposals such as an atomic airplane.) The Mesa Verde formation lay some 8,400 feet below ground, thus shielding the surface from any radioactive fallout. The government contributed public land at Battlement Mesa and 10 percent of the costs.[4]

The experiment succeeded in liberating gas, but it was (of course) radioactive and deemed unsuitable for household use. Even the energy crises that engulfed the country during the 1970s could not argue for the program, although subsequent wells, with nonnuclear technology, now dapple the slopes down to the town of Parachute. From above, the landscape looks like the high-tech version of a prairie dog town. The Department of Energy began cleanup operations from the explosion that continued into 1998. The State of Colorado established a buffer zone. A placard was erected in 1976.

That same year nature found ways to liberate more traditional combustibles. On July 11 lightning kindled a fire that flared into view the next day in Eames Orchard outside Parachute. By 5 p.m. the Grand Valley Volunteer Fire Department had controlled the fire at half an acre. On the 15th it apparently reignited, and this time with wind and cheatgrass to carry its spread upcanyon toward the tangle of gas wells, pipelines, and the newly plaqued Project Rulison site. These were still the early years of the fire revolution, in which interagency cooperation was promoted for fire suppression but the 10 a.m. rule remained as an inflexible principle. Still, there was added urgency to have crews and air tankers keep the fire away from those modern ancient fuels, even as those crews were transported and planes flown by distilled petroleum.

Of the three air tankers one, a B-26 piloted by Donald Goodman, crashed the next morning. The following day, during burnout operations along the ridgetop, three members of the Mormon Lake Hotshots were killed and a fourth badly burned. On the 18th rain fell. On the 19th the fire was declared controlled. On the 20th it was out. The final burn blackened 880 acres. The next year Carl Wilson published his famous study on common factors in fatality fires, which tend to be small, susceptible to sudden changes in relative humidity and wind, often in steep terrain, and happen during transitions. The Battlement Creek fit that profile nicely.[5]

And that is where the American wildland fire community has been content to have it reside. As with South Canyon it offers lessons in how to keep crews safe. Like the Eames Orchard reburn similar scenes keep recurring. The plaques, the memorial trails, the staff rides—these are the context for remembering the fire. Yet if you hover over Battlement Mesa today (or Google Earth it), you will see a landscape patchy with gas wells. Whatever its needs to manage wildland fire, the country has a deeper political commitment to keeping unimpeded the flow of combustion from fossil fuels. The South Cañon Number 1 coal mine continues to burn. In fact, some 35 coal mines in Colorado alone hold fire.

These fuels, and their associated fires, are also part of American fire history, or more broadly, of Earthly fire history. They speak to humanity as a fire creature, able to burn lithic landscapes as well as living ones. They remind us that our environmental power is fundamentally a fire power, that our industrial civilization is one founded on burning fossil fuels, and that these, too, can sometimes turn feral or lead to unhappy choices.

To date, we have no memorials toward our errors or tragic outcomes in nominally managing those burns. The alarm over global climate change, leveraged by combusting fossil fuels, is starting to alter that perspective, but it is not seen as a fire story. It is. Since the moment *Homo* first picked up a firestick, our decisions have been unsettling the planet. It's no longer good enough to pretend that fire history is a subset of climate history. Rather, climate history has become a subset of fire history. Our fires no longer interact only with air, and afterwards, with water, but with earth.

Viewed this way the fatality fires in Colorado can merge with all the other fatalities caused by our addiction to fossil fuels. Maclean's meditation embedded fire within the human condition. But our fire technologies are also a foundational part of the human condition. There are times on a fireline when you have to choose, and those choices can have lethal consequences. The same is true for a society, or since all human societies are desperate for more power, and find it in fire, for humanity. Our species history is a fire history.

That's a much broader frame than the wildland fire community is accustomed to, or even wants. It's probably a narrative that only an intellectual could love. It's good to have—good to be reminded that our firepower saturates our civilization. But it's also good to remember those names on the plaques at South Canyon and Battlement Creek. The fire community can't solve the problems of global warming, urban sprawl, invasive species shunted around the globe by oil-powered cargo ships and airplanes, and the toxic residues left from mines and wells. It has to look after its own. But sometimes it's good to recall that we are members of a much larger fire community that spans the globe and a nontrivial fraction of Earth time and that also knows fire disaster and fire tragedy without ever donning Nomex or hefting a pulaski.

EPILOGUE

The Interior West Between Two Fires

IN 1867 CLARENCE KING, then 25 years old, a graduate of Yale University's Sheffield Scientific School and a veteran of the Geological Survey of California, convinced the U.S. Army to appoint him director of a survey that would cross between the Sierra Nevada and the Rocky Mountains. The Geological Exploration of the 40th Parallel planned to roughly follow the route of the transcontinental railroad, then two years away from completion, and assess the natural resources on a 100-mile swath to either side of the rails.

It was a brilliant idea: steam power would link the country together, and it promised a rearrangement of human relations to nature across the Interior West. Humanity's new firepower would subordinate its old ones. The rails would bring in new capital, ideas, and people, and they would export the products of that alliance to markets throughout the globe. That dream, or salt-flat mirage, instead imported hordes of livestock, invasive weeds, and laissez-faire money that wrecked much of the natural estate. The vision of a new Comstock stumbled because ore was not renewable on anything like a human scale. (The closest approximation was an urban creation like Las Vegas that could keep reinventing ways to capture bullion.)

None of that was known at the time. When, in 1879 the King Survey, one of four in the West, was amalgamated with the others into the U.S. Geological Survey, the future of the Interior West seemed likely to be based on rock and water. Agriculture would mean irrigation and landscape-scale ranching, quite unlike anything in the east, but still an economy of commodities. Legal and political quarrels were over water rights and mineral claims, and later, over the abuse of the commons, the public domain. Clarence King was appointed the Survey's first director. Some 18 months later John Wesley Powell replaced him, and with that succession the Powell group's vision of a science-informed, state-sponsored program of conservation found institutional grounding.[1]

The centennial of the USGS saw a different West emerging, quickened by over a decade of bipartisan legislation on environmental matters. Geology had undergone a revolution, signified by the theory of plate tectonics, that characterized the Great Basin—the Basin Range province—as a zone of tension in which immense plates of planetary crust were slowly pulling apart. The Survey celebrated its hundredth anniversary by establishing a G. K. Gilbert Fellowship for outstanding work, thus honoring a member of the Powell group, its first chief geologist, and the man who worked out (and named) the geology behind the Basin Range. (Interestingly, Gilbert was responsible for a famous Washington luncheon group, a kind of Cosmos Club for field folk, which he named the Great Basin Mess in honor of his years in the region.)

That same era witnessed America's great cultural revolution on fire. In 1978, a century after the Powell Survey's map of Utah fires, the U.S. Forest Service formally replaced the 10 a.m. policy with a policy of fire by prescription. This, too, forced a reconsideration of the Great Basin, though it looked at ecological formations, not geologic ones; scrutinized the deep dynamics by which fire interacted with sagebrush steppes, pinyon-juniper woodlands, cheatgrass, and aspen; and reassessed the role of the federal government in overseeing the public lands. By now the economy had swerved from raw commodities into amenity communities, services, and urban metropoli. Partisans of the Sagebrush Rebellion argued to transfer most of the Interior West's public lands to the states or to private ownership. The environmental crisis lay less with the 1872 Mining Act than with the 1973 Endangered Species Act. Instead of a Geological Survey, a National Biological Survey (later, Service) was established to better address the assorted crises of the country's nonagricultural ecological estate. Meanwhile, a significant fraction of conservation had evolved into preservation, and fire protection had morphed into wildland fire management. Combustion as a generic source of power was challenged, at least conceptually, by a search for renewable energy in sun and wind.

In this way the Interior West has two pyrogeographies. One is the ancient realm of lightning and torch, bunchgrass and sagebrush, woodland and salt flat. Their regimes follow the primordial logic of fire since the Devonian. The ranges hold the frequent forest fires. The basins collect the grass and shrub and less frequent woodland fires. The other combustion realm burns lithic landscapes. The hump and hole that characterizes the region's gross physiography has a parallel in its pyrogeography. Symbolically, fire flows from the hump, Colorado, a source of fossil fuels, to the hole, Nevada, a neon-flashing sink for them. Utah has a bit of both, but mostly passes the flow to the West Coast.

The larger of the two realms of fire, as measured by fuel consumed and emissions, resides with lithic landscapes. It's hard to imagine much of an economy for the Interior West without fossil fuels, automobiles, air conditioners, and the city-states they sustain, which turn the wheels of the western economy. Secretary of the Interior for the Obama Administration, Sally Jewell, had run REI before accepting her appointment—recreation had more economic clout than forest and range commodities. Ranching and agriculture exist mostly through subsidies, however potently they persist as imagined worlds. Cities make the economy, and combusted energy powers the cities. Its two urban clusters account for 90 percent of Nevada's population, the Wasatch Front for over 70 percent of Utah's, and the Front Range 89 percent of Colorado's. They ensure that, over time, urban values will triumph. Yet rural America loosens its grip over national politics and the national psyche grudgingly, however far it ranges from the national economy. It is here that landscape fire meets city fringes, so it is here that most of the fire establishment concentrates.

On May 10, 1869, the two lines of the transcontinental railway joined at Promontory Summit in northern Utah. Despite the efforts of the Geological Exploration of the 40th Parallel to find riches for the rails to carry, most traffic and people just sought to pass through. Only the Mormon search for sanctuary in a deliberate effort to avoid the kinds of places that attracted Gentiles created anything permanent between the Front Range and the Sierra Nevada. Both of Nevada's urban clusters remain tentacled to California.

So, too, the national fire establishment has largely passed over the Interior West. Federal agencies bonded the region to the rest of the country, even if, as with rail and later highway, they mostly moved fire traffic across the land. The Bureau of Land Management made the Great Basin the heartland for its operations in the Lower 48, but fires were either sparse or, after cheatgrass began feeding serious conflagrations, still beyond the pale of national interest. For a region that so long relied on minerals, it will be paradoxical if its ecology in the form of the greater sage grouse becomes the golden spike that links lithic and living landscapes and joins the Interior West to the national narrative of fire. The transcontinental railroad revolutionized nature's economy as well as humanity's. What the region's new firepower brings will likely decide whether the center holds or splinters off, to be annexed by the powerful regions around it. The Arid Regions were once an inspiration for a national program of conservation. Perhaps they can again pioneer a new era of fire management.

Or maybe, what seems most likely, the Interior West will be neither core nor periphery, neither hole nor exoskeleton, but simply part of the indissoluble matrix that constitutes fire in America. The United States holds nearly a billion burnable acres. A significant fraction lies in the Interior West. Those fires now fill what two centuries ago had seemed a pyric emptiness. The national narrative can no longer simply pass over the region with hardly a pause because the regional story—the arid interior with its vulnerable sage steppes, its awkward pinyon-juniper woodlands, its maddening cheat-infested grasslands, its humble aspen—has become a vital subplot without which the national one can no longer make sense. The center doesn't have to hold the rest. It just has to be there.

NOTE ON SOURCES

AS USUAL THIS project was an exercise in encountering new landscapes (or old landscapes in a new way) and some of the people who must manage fire in them. The written literature was background to what those encounters prompted. The notes provide the names of both people and books. But much of the pleasure in a reconnaissance of the Interior West was the chance to explore the classic literature, including historical. A handful of those works are worth repeating here.

Some were old favorites, such as Powell's *Report on the Lands of the Arid Region of the United States* and Wallace Stegner's *Mormon Country*, which also provided a loose model for interpreting a region through select essays. A few were venerable histories such as Leonard Arrington's *Great Basin Kingdom* and Richard Lillard's *Desert Challenge*. And some were fresher; I found particularly helpful James Young and Abbott Sparks's *Cattle in the Cold Desert* and Thomas Noel's *Colorado: A Historical Atlas*. The newer histories are so urban, and in the case of Nevada, so consumed by Las Vegas, that they were less useful, though informative and often entertaining. Because so much of the land is public domain, there is a vast scientific and technical literature published primarily through symposia, journals, and the Forest Service. The most relevant of these I identify in the notes specific to their topic and place.

NOTES

PROLOGUE: ARID LANDS, BURNING LANDS

1. The classic study of the conservation movement remains, in many ways, the best. Samuel P. Hays, *Conservation and the Gospel of Efficiency: The Progressive Conservation Movement, 1890–1920* (Cambridge, MA: Harvard University Press, 1959).
2. J. W. Powell, *Report on the Lands of the Arid Region of the United States* (Washington, DC: Government Printing Office, 1878; repr., 1879), 15.
3. Powell, *Report*, 15.
4. Powell, *Report*, 17.
5. Powell, *Report*, 17–18.
6. Powell, *Report*, 24, 18.
7. J. W. Powell, "The Non-Irrigable Lands of the Arid Region," *Century Magazine* 39 (1890): 919.
8. Quote from B. H. Roberts, *A Comprehensive History of the Church of Jesus Christ of Latter-Day Saints*, but by way of Sara Dant, "Field Notes: Brigham Young's 'All the People' Quote Quandary," *Western Historical Quarterly*, 46, no. 2 (Summer 2015): 219–23. I'm grateful to Sara for sharing an in-publication, more expansive version that prodded me to rethink how to interpret the environmental history of the Plateau Province; see Sara Dant, "The 'Lion of the Lord' and the Land: Brigham Young's Environmental Ethic," in *The Earth Will Appear as the Garden of Eden: Essays on Mormon Environmental History*, ed. Jedidiah Rogers and Matthew C. Godfrey (Salt Lake City: University of Utah Press, 2018).

9. Powell, "Non-Irrigable Lands," 919. For Fernow's reaction, see Andrew Denny Rodgers III, *Bernhard Eduard Fernow: A Story of North American Forestry* (New York: Hafner, 1968), 167.
10. Ibid.
11. F. E. Clements, *The Life History of Lodgepole Burn Forests*, Bulletin 79, U.S. Forest Service, 1910.

NEVADA: FROM ROTTEN BOROUGH TO BURNING MAN

1. One of the pleasures of exploring Nevada are the wonderful histories of its first century. I particularly enjoyed Gilman M. Ostrander, *Nevada: The Great Rotten Borough: 1859–1964* (New York: Knopf, 1966); Richard G. Lillard, *Desert Challenge: An Interpretation of Nevada* (New York: Knopf, 1942); and Dale L. Morgan, *The Humboldt: Highroad of the West* (New York: Farrar and Rinehart, 1943). Useful as a concise survey was Michael W. Bowers, *The Sagebrush State: Nevada's History, Government, and Politics* (Reno: University of Nevada Press, 1996). An interesting view of the Great Basin and its urban clusters is Dennis R. Judd and Stephanie L. Witt, eds., *Cities, Sagebrush, and Solitude: Urbanization and Cultural Conflict in the Great Basin* (Reno: University of Nevada Press, 2015).
2. Wallace Stegner, *Mormon Country* (1942; repr., Lincoln: University of Nebraska Press, 1981): 45.
3. Letter to Jane Clemens, 1861; Mark Twain, *The Works of Mark Twain*, vol. 2, *Roughing It* (1872; repr., Berkeley: University of California Press, 1972): 15. Quote on Great Basin from Great Basin Restoration Initiative, *Out of Ashes, An Opportunity* (Boise, ID: Bureau of Land Management, 1999): 4.
4. Mike Stambaugh, "Wave of Fire: Fire History and Settlement Patterns Across the Eastern U.S." (presentation, Organization of American Historians annual meeting, St. Louis, MO, 2015).
5. Walter Van Tilburg Clark, *The Track of the Cat* (New York: New American Library, 1949): 14.
6. See Shawna Legarza, *No Grass* (n.p.: BookSurge Publishing, 2009).

THE OTHER SIDE OF THE MOUNTAIN: WASHOE WUI

1. A lot of people to thank for contributing their time and knowledge. For the Forest Service, Mike Wilde and Mike Dondero. For the BLM,

Tim Roide, Keith Barker, Ryan Elliott, Dennis Terry, Billy Britt, and for a broader overview, Paul Petersen. From NDF, Joe Freeland, who gathered most of his fire staff for a vigorous conversation.

2. Mark Twain, *Roughing It* (New York: New American Library, 1962): 127–28.
3. The basic history of the NDF is available through its website, http://forestry.nv.gov, through links to "State Fire Warden Position" and "Division of Forestry."
4. The latest (June 7, 2016) of the strategic plan is available online at http://forestry.nv.gov, accessed December 3, 2016. Quote from page 2.

A SINK FOR EXOTICS

1. Aldo Leopold, "Cheat Takes Over," in *A Sand County Almanac* (New York: Ballantine Books, 1970): 165. James A. Young and Charlie D. Clements, *Cheatgrass: Fire and Forage on the Range* (Reno: University of Nevada Press, 2009): xiii. I have relied on this fascinating volume for my background understanding of cheatgrass. For historical background see James A. Young and B. Abbott Sparks, *Cattle in the Cold Desert* (Logan: Utah State University, 1985). To put cheatgrass within the national spectrum of invasives, see Kristin Zouhar et al., *Wildland Fire in Ecosystems: Fire and Nonnative Invasive Plants*, General Technical Report RMRS-GTR-42-vol. 6, U.S. Forest Service, 2008.
2. Young and Clements, *Cheatgrass*, xiii–xiv.
3. Many references in the literature, but the best quantitative summary is Jennifer K. Balch et al., "Introduced Annual Grass Increases Regional Fire Activity Across the Arid Western USA (1980–2009)," *Global Change Biology* 19 (2013): 173–83.
4. My two primary sources for this topic are Young and Clements, *Cheatgrass*, and Ronald Tobey, *Saving the Prairies: The Life Cycle of the Founding School of American Plant Ecology, 1895–1955* (Berkeley: University of California Press, 1982). The analogy of ideas to invasives is, alas, my own.
5. G. D. Pickford, "The Influence of Continued Heavy Grazing and the Promiscuous Burning on Spring-Fall Ranges in Utah," *Ecology* 13 (1932): 159–71.
6. See, for example, Bruce L. Welch, *Big Sagebrush: A Sea Fragmented into Lakes, Ponds, and Puddles*, General Technical Report RMRS-GTR-144, U.S. Forest Service, 2005.
7. Leopold, "Cheat Takes Over," 167–68.

A WORTHY ADVERSARY

1. Biographical material from Mike Pellant, mostly obtained by a phone interview on September 26, 2016.

   Other useful documents include his resume; Pellant's statement before a Senate subcommittee, at https://www.blm.gov/sites/blm.gov/files/congressional_testimony_documents/congressional_20071011_RegardingtheGreatBasinRestorationInitiative.pdf; and Mike Pellant, "Reflections on 30+ Years of Tackling the Cheatgrass/Wildfire Cycle in the Great Basin," accessed September 29, 2016, https://environment.unr.edu/consortium/downloads/Presentations13/early-powerpoint-version-pellant-gbc-plenary-1-14-13.pdf.
2. Mike Pellant, "History and Applications of the Intermountain Greenstripping Program," in *Proceedings—Ecology and Management of Annual Rangelands* (Boise, ID: Intermountain Research Station, 1994): 63–68.
3. Mike Pellant, "Cheatgrass: The Invader That Won the West" (presentation, Interior Columbia Basin Ecosystem Management Project, 1996).
4. Pellant, "Cheatgrass," 13–14.
5. Statistics from BLM, *Out of Ashes, an Opportunity*, 6.

MUSHROOM CLOUDS

1. William J. Broad, "The Hiroshima Mushroom Cloud That Wasn't," *New York Times*, May 23, 2016, http://www.nytimes.com/2016/05/24/science/hiroshima-atomic-bomb-mushroom-cloud.html. To clarify, quote is from article, not Roark.
2. Horatio Bond, *Fire and the Air War* (Boston, MA: National Fire Protection Association, 1946).
3. The most concise summary of events is available in Stephen J. Pyne, *Fire in America* (Seattle: University of Washington Press, 1999): 480–90; Brown quote on 480–81.
4. See Pyne, *Fire in America*, 487.
5. A detailed exploration of this topic is Lynn Eden, *Whole World on Fire: Organizations, Knowledge, and Nuclear Weapons Devastation* (Ithaca, NY: Cornell University Press, 2004).
6. The image of Libby was published in several venues and has been reproduced. A version can be found online in the Jasper, Indiana, *Herald* (October 5, 1961), 19. The Nixon photo by Allan Grant first appeared in the Los Angeles *Times*, and was then republished in *Life*. An online ver-

sion is available in an obituary of Grant, published by the *Times*: http://www.latimes.com/local/obituaries/la-me-allan_grant-pg-photogallery.html.

7. Background fire ecology available in Matthew L. Brooks, Todd C. Esque, and Tim Duck, "Creosotebush, Blackbrush, and Interior Chaparral Shrublands," chap. 6 in General Technical Report RMRS-GTR-202, U.S. Forest Service, 2007.

8. Janice C. Beatley, "Ecological Status of Introduced Brome Grasses (*Bromus spp.*) in Desert Vegetation of Southern Nevada," *Ecology* 47, no. 4 (1966): 548–54. For background on her plots, see Robert H. Webb et al., "Monitoring of Ecosystem Dynamics in the Mojave Desert: The Beatley Permanent Plots," USGS Fact Sheet FS-040-01, April 2001. The NEPA study issued annual reports from 1987 to 1994 under the title *Status of the Flora and Fauna on the Nevada Test Site, 1994: Results of Continuing Basic Environmental Monitoring January Through December 1994* (the years change, the rest of the title remains the same). These are available through the Office of Scientific and Technical Information Scitech Connect website: http://www.osti.gov/scitech.

   I would like to thank Martha DeMarre of Nuclear Testing Archive, National Security Technologies, a contractor to DOE, for assistance in finding the relevant reports.

9. The compliance reports are available at the OSTI website: www.osti.gov/scitech.

10. Dennis J. Hansen and W. Kent Ostler, "A Survey of Vegetation and Wildland Fire Hazards on the Nevada Test Site," DOE/NV/25946-083(2006); quotes from 1, 4.

DEEP FIRE

1. In keeping with its 10,000-year horizon, the Long Now Foundation uses a five-digit timeline. It seemed appropriate to keep that convention for this essay.

2. Information on the Phillips Ranch fire come from Alexander Rose of the Long Now Foundation in an e-mail dated July 25, 2016, which also included e-mails between Stewart Brand and Rebecca Mills, superintendent of Great Basin National Park, and several press releases.

3. I'd like to thank Matt Martin for help in setting up a site visit, although we could not coordinate our calendars, and Matt Johnson for sending me copies of the park's fire management plan, fire atlas, and fire statistics.

4. Brand quote from Long Now Foundation website at http://longnow.org/clock/nevada/, accessed July 25, 2016.
5. Many accounts of these changes are in print, at various levels of detail. A useful thumbnail is available in Donald K. Grayson, "Great Basin Natural History," in Catherine S. Fowler and Don D. Fowler, eds., *The Great Basin: People and Place in Ancient Times* (Santa Fe, NM: School for Advanced Research Press, 2008): 7–17.
6. The standard history of the range, cave, and park is Darwin Lambert, *Great Basin Drama: The Story of a National Park* (Niwot, CO: Rinehart, 1991), which is especially good on the politics.
7. Stanley G. Kitchen, "Historical Fire Regime and Forest Variability on Two Eastern Great Basin Fire-Sheds (USA)," *Forest Ecology and Management* 285 (2012): 53–66, quoted line from 53.
8. Emily K. Heyerdahl, et al., "Multicentury Fire and Forest Histories at 19 Sites in Utah and Eastern Nevada," General Technical Report RMRS-GTR-261WWW, U.S. Forest Service, 2011; and George E. Gruell, "Historical and Modern Roles of Fire in Pinyon-Juniper," in Stephen Monsen and Richard Stevens, comps., *Proceedings: Ecology and Management of Pinyon-Juniper Communities Within the Interior West*, Proceedings RMRS-P-9 (Ogden, UT: U.S. Forest Service, 1999): 24–28.
9. Gruell, "Historical and Modern Roles," 26. On the Strawberry fire: http://inciweb.nwcg.gov/incident/4947/, accessed September 2, 2016. On general park statistics, see Great Basin National Park, *Fire Management Plan*, n.d., 12–13.
10. A delightful popular introduction is Ronald M. Lanner, *The Bristlecone Book: A Natural History of the World's Oldest Trees* (Missoula, MT: Mountain Press, 2007). A meditation, particularly useful in teasing out the history of progressive understandings, is Michael P. Cohen, *A Garden of Bristlecones: Tales of Change in the Great Basin* (Reno: University of Nevada Press, 1998). For basic fire ecology, see Fire Effects Information System at https://www.feis-crs.org/feis/. Also tantalizing is a poster at a fire ecology conference by Mackenzie Kilpatrick and Franco Biondi, "Bristlecone Fire History and Stand Dynamics at Mount Washington, Nevada," which looked at the aftermath of the 2000 fire.
11. On the clock see the Long Now Foundation's wonderful website at http://longnow.org/essays/time-10000-year-clock/, accessed August 1, 2016.
12. I follow Cohen's matrix of interpretations in *Garden of Bristlecones*, 61–77. The number of conflicting accounts is remarkable. Currey quoted

from "An Ancient Bristlecone Pine Stand in Eastern Nevada," *Ecology* 46 (1965): 566. Contrast Currey's bland, neutral-voice article with Lambert's in *Great Basin Drama*, 142–47, which includes photos of the tree before felling (and Currey climbing it).

OUTLIER: STRIP TRIP

1. Clarence Dutton, *Tertiary History of the Grand Cañon District*, U.S. Geological Survey Monograph 2 (Washington, DC: Government Printing Office, 1882), 78–79.
2. A brief survey is available at the BLM Arizona Strip website: http://www.blm.gov/az/st/en/prog/blm_special_areas/natmon/gcp/cultural/ranching.html, accessed June 6, 2016. An unpublished history compiled by Phil Foremaster, "History, Geology, and Ranching on the Arizona Strip North of the Colorado," provides considerable details up through 1934 (see website of the Washington County Historical Society, http://wchsutah.org/az-strip/history-geology-ranching.pdf).
3. Bob Davis, "The History of the Mt. Trumbull Forest (As I Know It)," unpublished report on file at BLM Arizona Strip District Office.
4. Tree density numbers from Ken Moore, Bob Davis, and Timothy Duck, "Mt. Trumbull Ponderosa Pine Ecosystem Restoration Project," in *Visitor Use Density and Wilderness Experience: Proceedings*, RMRS-P-20 (Missoula, MT: U.S. Forest Service, 2003): 120.
5. A master bibliography of published and unpublished reports is available on the project website: http://nau.edu/eri/research/ecological-research/arizona/mount-trumbull/.

   For a textual synthesis see David W. Huffman et al., "Ecosystem Restoration, Assistance Agreement No. PAA 017002, Task Order No. AAW040001, Final Report, February 5, 2006." Excellent digests can be found in Moore, Davis, and Duck, "Mt. Trumbull Ponderosa Pine," and Thomas Heinlein et al., "Changes in Ponderosa Pine Forests of the Mt. Trumbull Wilderness," unpublished report to BLM (November 1999). The best summary, and by far the most readable, is Peter Friederici, "Healing the Region of Pines: Forest Restoration in Arizona's Uinkaret Mountains," in Peter Friederici, ed., *Ecological Restoration of Southwestern Ponderosa Pine Forests* (Washington, DC: Island Press, 2003): 197–214.
6. For a good synopsis see "Restoring the Uinkaret Mountains: Operational Lessons and Adaptive Management Practices," Working Papers

in Southwestern Ponderosa Pine Forest Restoration 1, Ecological Restoration Institute (n.d.).

7. Friederici, "Healing the Region of Pines," 204–5, 213.
8. Dutton, *Tertiary History*, 92. Holmes's atlas art can be viewed at https://www.loc.gov/resource/g4332gm.gnp00002/?st=gallery. See sheet VI for Toroweap and sheets XV–XVII for Point Sublime.
9. Dutton, *Tertiary History*, 89.

UTAH: ZION'S HEARTH

1. Dale L. Morgan, *The State of Deseret* (Logan, UT: Utah State University Press and Utah Historical Society, 1987); Leonard Arrington, *Great Basin Kingdom: Economic History of the Latter-day Saints, 1830–1900* (Lincoln: University of Nebraska Press, 1966); Richard V. Francaviglia, *The Mormon Landscape* (New York: AMS Press, 1978); Donald Meinig, "The Mormon Culture Region: Strategies and Patterns in the Geography of the American West, 1847–1964," *Annals of the Association of American Geographers* 55, no. 2 (June 1965): 191–220.
2. See Gary Topping, *Utah Historians and the Reconstruction of Western History* (Norman: University of Oklahoma Press, 2003). For a sympathetic survey of Mormon, not just Utah, literature, see Terryl L. Givens, *People of Paradox: A History of Mormon Culture* (Oxford: Oxford University Press, 2007): 285–324.
3. Writing in 1942 Wallace Stegner thought it "almost impossible to write fiction about the Mormons, for the reason that Mormon institutions and Mormon society are so peculiar that they call for constant explanation. The result is local color, an almost unavoidable leaning toward the picturesque qualities of a unique social order, and the Mormons as people get lost behind the institutional barriers that set them apart. Still, it is the sticks which furnish the best material for historical best sellers, and the Mormon Country will be the sticks for a good while to come." Stegner, *Mormon Country* (Lincoln: University of Nebraska Press, 1970): 347–48.

CATFIRE: WASATCH WUI

1. Thanks to Matt Snyder (and others), Brad Washa, Pat Shea, Dan Washington, and Lester Brown for an introduction to fire and life along the Wasatch Front, and beyond.

2. Rebecca Andersen, "Between Mountain and Lake: an Urban Mormon Country" (PhD diss., Arizona State University, 2015). I'm indebted to Rebecca for informing me about the history of Mormon urban life and for the history of Envision Utah.
3. Utah Department of Agriculture and Food, *Catastrophic Wildfire Reduction Strategy: Protecting the Health and Welfare of Utahans and Our Lands*, 2014.
4. An old but still pertinent study is Walter P. Cottam and Frederick R. Evans, "A Comparative Study of the Vegetation of Grazed and Ungrazed Canyons of the Wasatch Range, Utah," *Ecology* 26, no. 2 (April 1945): 171–81.
5. See Cottam and Evans, "A Comparative Study," and James R. Ehleringer et al., "Red Butte Canyon Research Natural Area: History, Flora, Geology, Climate, and Ecology," *Great Basin Naturalist* 52, no. 2 (June 1992): 95–121.
6. See Utah DNR, "Communities at Risk to Wildland Fire," http://www.arcgis.com/home/webmap/viewer.html?webmap=1344a0880bb1444baf7332b58e7276fb&extent=-116.8964,36.1207,-106.1298,42.812.

   On the state's 2015 fire history, see https://www.arcgis.com/home/item.html?id=3e926e4e18e54359a513407d7c56fadc, accessed July 20, 2015.

BURNING BUSHES

1. Literature is expanding as rapidly as pinyon-juniper throughout the West. I found Ronald M. Lanner's *The Piñon Pine: A Natural and Cultural History* (Reno: University of Nevada Press, 1981) to be a sweet introduction. Otherwise, I relied on the proceedings of three conferences: Richard L. Everett, comp., *Proceedings: Pinyon-Juniper Conference, Reno, NV, January 13–16, 1986*, General Technical Report INT-215, U.S. Forest Service, 1987; Douglas W. Shaw, Earl F. Aldon, and Carol LoSapio, tech. coords., *Desired Future Conditions for Piñon-Juniper Ecosystems, August 8–12, 1994, Flagstaff, Arizona*, General Technical Report RM-259, U.S. Forest Service, 1995; and Stephen B. Monsen and Richard Stevens, comps., *Proceedings: Ecology and Management of Pinyon-Juniper Communities Within the Interior West*, Proceedings RMRS-P-9, U.S. Forest Service, 1999. A good survey of current sentiments is available in W. C Romme et al., *Historical and Modern Disturbance Regimes of Piñon-Juniper Vegetation in the Western US*, Colorado Forest Restoration Institute, Colorado State University, 2008, at https://cfri.colostate.edu.

2. Richard F. Miller and Peter E. Wigand, "Holocene Changes in Semiarid Pinyon-Juniper Woodlands," *BioScience* 44, no. 7 (July/August 1994): 465–74, quote from 468. A good sense of the views as they existed in the late 1970s is available in Garry F. Rogers, *Then and Now: A Photographic History of Vegetation Change in the Central Great Basin Desert* (Salt Lake City: University of Utah Press, 1982).
3. On a general model of aboriginal burning, see Stephen J. Pyne, *Fire: A Brief History* (Seattle: University of Washington Press, 2001) and "Fire Creature," in Andrew Scott et al., *Fire on Earth: An Introduction* (Chichester, West Sussex: Wiley Blackwell, 2014).
4. For the workshop's conclusions, see Romme et al., *Historical and Modern Disturbance Regimes*. A short version was published as "Historical and Modern Disturbance Regimes, Stand Structures, and Landscape Dynamics in Piñon-Juniper Vegetation of the Western United States: Invited Synthesis Paper," *Rangeland Ecology and Management* 62, no. 3 (May 2009): 203–22.

PLATEAU PROVINCE

1. I have benefited hugely from conversations with many people in the region. Let me here thank those who were especially helpful. Zion: Taiga Rohrer. Henry Mountains: Doug Page, Charles Kay, Todd Murray. High Plateaus: Linda Chappell, Kevin Greenhalgh, Gayle Sorensen, Tyler Monroe, Russ Ivey, Glen Chappell.
2. Sharon M. Hood and Melanie Miller, eds., *Fire Ecology and Management of the Major Ecosystems of Southern Utah*, General Technical Report RMRS-GTR-202, U.S. Forest Service, 2007, 1.
3. David A. Prevedel and Curtis M. Johnson, *Beginnings of Range Management: Albert F. Potter, First Chief of Grazing, U.S. Forest Service, and a Photographic Comparison of His 1902 Forest Reserve Survey in Utah with Conditions 100 Years Later*, R4-VM 2005-01 (Ogden, UT: Intermountain Region, U.S. Forest Service, 2005); David A. Prevedel, E. Durant McArthur, and Curtis M. Johnson, *Beginnings of Range Management: An Anthology of the Sampson-Ellison Photo Plots (1913–2003) and a Short History of the Great Basin Experiment Station*, General Technical Report RMRS-GTR-154, U.S. Forest Service, 2005.
4. For the regional setting, see Emily K. Heyerdahl et al., *Multicentury Fire and Forest Histories at 19 Sites in Utah and Eastern Nevada*, General Technical Report RMRS-GTR-261WWW, U.S. Forest Service, December 2011.

5. The diminution of aspen has been a research theme since the 1980s. Useful distillations are available in Dale L. Bartos and Robert B. Campbell Jr., "Decline of Quaking Aspen in the Interior West—Examples from Utah," *Rangelands* 20, no. 1 (February 1998): 17–24. A useful survey based on historical records is available in Karen Ogle and Valerie DuMond, *Historical Vegetation on National Forest Lands in the Intermountain Region* ([Ogden, UT?]: U.S. Forest Service, Intermountain Region, 1997). Pair it with Charles E. Kay, *Long-Term Vegetation Change on Utah's Fishlake National Forest: A Study in Repeat Photography* (August 2003), at http://idahoforwildlife.com/files/pdf/charlesKay/55-%20Long-Term%20Vegetation%20Change%20on%20Utah%20%20Fishlake%20National%20Forest%20A%20Study%20in%20Repeat%20Photography.pdf. More broadly, see Daniel deB. Richter, "The Accrual of Land Use History in Utah's Forest Carbon Cycle," *Environmental History* 14 (July 2009): 527–42, which attempts to quantify some of the processes involved.
6. Frederick Baker, *Aspen in the Central Rocky Mountain Region*, Bulletin No. 1291, U.S. Department of Agriculture, 1925, 19.
7. See, for example, Charles E. Kay, *Aspen Management Guidelines for BLM Lands in North-Central Nevada: Final Report to Battle Mountain Field Office* (Battle Mountain, NV: Battle Mountain Field Office, Bureau of Land Management, 2003), 1. As with so much of the region, grazing is the critical intermediary in fire history. On Pando, see Michael C. Grant, "The Trembling Giant," *Discover* (October 01, 1993); online access at http://discovermagazine.com/1993/oct/thetremblinggian285.
8. See, for example, James N. Long and Karen Mock, "Changing Perspectives on Regeneration Ecology and Genetic Diversity in Western Quaking Aspen: Implications for Silviculture," *Canadian Journal of Forest Research* 42 (2012): 2011–21.
9. See Michael H. Madany and Neil E. West, "Fire History of Two Montane Forest Areas of Zion National Park," in Marvin A. Stokes and John H. Dieterich, tech. coords., *Proceedings of the Fire History Workshop, October 20–24, 1980, Tucson, Arizona*, General Technical Report RM-81, U.S. Forest Service, 1980, 50–56; and Michael H. Madany and Neil E. West, "Livestock Grazing-Fire Regime Interactions Within Montane Forests of Zion National Park, Utah," *Ecology* 64, no. 4 (1983): 661–67.
10. Madany and West, "Fire History," 56. For a concise summary of Zion's fire history, see *Zion Fire Facts*, https://home.nps.gov/zion/learn/nature/upload/ZION-Fire-Facts-2012-FINAL.pdf. Regarding the vexing problem of cheatgrass, see Kelly Fuhrmann, "Restoring Burned Area Fire Regimes at Zion National Park," in *Proceedings of the 2007 George*

Wright Society Conference (Hancock, MI: George Wright Society, 2008), 189–93. The park resorted to herbicides and reseeding to try to prevent cheatgrass reclamation.

11. For a summary, see Omer C. Stewart, *Forgotten Fires: Native Americans and the Transient Wilderness* (Norman: University of Oklahoma Press, 2002), 245–46.
12. Information from *Mill Flat Fire Review, March 2010*, Dixie National Forest, available online through the Wildland Fire Lessons Learned Center at http://www.wildfirelessons.net/HigherLogic/System/Download DocumentFile.ashx?DocumentFileKey=e9267f02-4fa2-478a-a057 -8ea5d1313fa2, accessed, November 10, 2016.
13. *Saddle Fire, Dixie National Forest, Timeline of Significant Events and Associated Decisions*, Dixie National Forest, July 22, 2016. For the final tally see the Inciweb report at http://inciweb.nwcg.gov/incident/4794/. Writing about recent events is always fraught. Participants disagree over why a NIMO team was ordered. I've bowed to the written documentation, with the understanding that those on the ground had a different perspective. What interests me is the striking contrast between the responses to the Mill Flat and Saddle fires.
14. See Charles E. Kay, *Long-Term Vegetation Change in Utah's Henry Mountains: A Study in Repeat Photography*, BLM-UT-GI-14001-8002, Bureau of Land Management, March 2014, and Heyerdahl et al., *Multicentury Fire*, "Appendix A: Henry Mountains, Bureau of Land Management (HMR)." For a detailed summary of settlement history, see Charles B. Hunt, *Geology and Geography of the Henry Mountains Region, Utah: A Survey and Restudy of One of the Classic Areas in Geology*, Geological Survey Professional Paper 228, U.S. Government Printing Office, 1953, 13–21. Supplement with Chas B. Hunt, "Around the Henry Mountains with Charlie Hanks," *Utah Geology* 4, no. 2 (Fall 1977): 95–104. I'm also grateful for e-mail exchanges with Stanley Kitchen regarding fire in the Henrys.
15. Stegner, *Mormon Country*, 349.

OUTLIER: MESA NEGRA

1. Acknowledgements. I want to thank Steven Underwood for helping to set up a site visit, and Keith Krause and the Mesa Verde engine crew for a delightful and informative study tour. I think it fair to say that how they see fire at Mesa Verde and how I see it differ, but then people have been seeing fire in various ways for a very long time on the cuesta.

2. Mesa Verde is unusually well endowed with fire history records. The best source is an unpublished report in the files of the Fire Management Office, *Mesa Verde Fire History, March 2002*, which gives detailed accounts of fires, a bibliography of published and unpublished sources, and even archival references. A popular version was published as part of the Mesa Verde Centennial Series: Tracey L. Chavis and William R. Morris, *Fire on the Mesa* (Durango, CO: Durango Herald Small Press, 2006).
3. Background fire ecology and fire history research is best distilled in M. Lisa Floyd, William H. Romme, and David D. Hanna, "Fire History and Vegetation Pattern in Mesa Verde National Park, Colorado, USA," *Ecological Applications* 10, no. 6 (2000): 1666–80, and a chapter by the same authors, "Fire History," in M. Lisa Floyd, ed., *Ancient Piñon-Juniper Woodlands: A Natural History of Mesa Verde Country* (Boulder: University Press of Colorado, 2003), 261–78. For criticisms of prescribed fire and mechanical treatments, see chapters by William H. Romme, Sylvia Oliva, and M. Lisa Floyd, "Threats to the Piñon-Juniper Woodlands," 339–61, and George L. San Miguel, "Epilogue: Management Considerations for Conserving Old-Growth Piñon-Juniper Woodlands," 361–74, in the same volume.
4. Conversations with Keith Krause, assistant fire management officer, and Mesa Verde Fire Management Plan.
5. On fire impacts, see Julie Bell, "Fire and Archaeology on Mesa Verde," in David Grant Noble, ed., *The Mesa Verde World: Explorations in Ancestral Pueblo Archaeology* (Santa Fe, NM: School of American Research Press, 2006), 119–21. See also Chavis and Morris, *Fire on the Mesa*, 99–100.
6. Noble, *Mesa Verde World*, 6.
7. On the fires of abandonment, see Noble, *Mesa Verde World*, 22, 26, 131, 129, 143. The savage depopulation is well documented throughout the New World. An interesting effort to reconcile that unloading with fire history is Matt Liebmann et al., "Native American Depopulation, Reforestation, and Fire Regimes in the Southwest United States, 1492–1900 CE," *PNAS* 113, no. 6 (2016), published online at http://www.pnas.org/cgi/doi/10.1073/pnas.1521744113.

COLORADO: ROCKY MOUNTAIN HIGHS, AND LOWS

1. I have found Thomas J. Noel, *Colorado: A Historical Atlas* (Norman: University of Oklahoma Press, 2015), invaluable as a source of information; employment statistics from page 130.

COLFIRE: THE FRONT AS CENTER

1. I wish to thank Mike Morgan, Vaughn Jones, and Caley Fisher for an informative tutorial on CDFPC, and Kristin Garrison for a similar introduction to CSFS. Both furnished copies of critical documents, which assisted my education considerably.
2. Most of the fires have been researched, as follows: National Fire Protection Association, *Black Tiger Fire Case Study* (Quincy, MA: NFPA, 1989). Russell T. Graham, tech. ed., *Hayman Fire Case Study: Summary*, General Technical Report RMRS-GTR-115, U.S. Forest Service, 2003, and for the full version, *Hayman Fire Case Study*, General Technical Report RMRS-GTR-114, U.S. Forest Service, 2003. Russell Graham et al., *Fourmile Canyon Fire Findings*, General Technical Report RMRS-GTR-289, U.S. Forest Service, 2012. William Bass et al., *Lower North Fork Prescribed Fire, Prescribed Fire Review, April 13, 2012*, report to Colorado Department of Natural Resources from Office of the President, Colorado State University. Inciweb, "High Park Fire," http://inciweb.nwcg.gov/incident/2904/. Waldo Canyon fire: Inciweb, "Waldo Canyon Fire," http://inciweb.nwcg.gov/incident/2929/; Stephen Quarles, et al., *Lessons Learned from Waldo Canyon: Fire Adapted Communities Mitigation Assessment Team Findings*; Alexander Maranghides et al., *A Case Study of a Community Affected by the Waldo Fire—Event Timeline and Defensive Actions*, NIST Technical Note 1910, National Institute of Standards and Technology, November 2015. Inciweb, "Black Forest Fire," http://inciweb.nwcg.gov/incident/3424/. For the background on the escalating scene, see CSFS, "Colorado Wildfires: State and Private Lands," at http://static.colostate.edu/client-files/csfs/documents/WILDFIRES_year_cb_2009.pdf.
3. Roxane White and Mark Gill, cochairs, *Review and Recommendations: Enhancing Fire Response and Management in Colorado State Government*, April 23, 2012. The sequence: the Lower North Fork fire occurred in March; the review committee submitted its report in April; the legislature responded in May and June, effective in July.
4. A dandy historical digest is available in White and Gill, *Review and Recommendations*, 9–10.
5. See CSFS, "Colorado Wildfires, State and Private Lands."
6. Two documents are particularly clarifying: Paul L. Cooke, et al., "Wildland Urban Interface Issues—A Fire Service Perspective," presented to Interim Committee on Wildfire Issues in Wildland-Urban Interface

Areas, August 6, 2008; and Colorado State Fire Chiefs' Association, "Fire Point Plan for a Fire Safe Colorado," n.d. I'm grateful to Caley Fisher for providing copies.

7. For a useful summary of the events that led to the transfer, see *Colorado Fire News* (November 1, 2012).
8. On treatments, see wildfire mitigation strategy, CSFS website, http://csfs.colostate.edu/wildfire-mitigation/.
9. Bryan Karchut, quote from e-mail message to author, August 12, 2016.

FIREBUGS

1. Beetles were part of the agenda when I met with some fire staffers at the Arapaho-Roosevelt National Forest. They passed along useful information and insights, and cannot in any way be implicated in how I have extrapolated from that discussion. But let me thank them again: Bryan Karchut, James White, Tim Haas, and Chet Dieringer. Poor Chet has the dubious distinction of being the only person I have met with twice in my research on *To the Last Smoke*. The last time was in east Texas.

   For an interesting account of a long-term effort to grapple with beetles, see Russell T. Graham et al., *Mountain Pine Beetles: A Century of Knowledge, Control Attempts, and Impacts Central to the Black Hills*, General Technical Report RMRS-GTR-353, U.S. Forest Service, 2016.
2. See Monique E. Rocca et al., "Did Prior Mountain Pine Beetle Activity Influence Patterns of Burn Severity in the 2012 High Park Fire, CO?" poster at 2015 meeting of Association for Fire Ecology, Large Fire Conference, San Antonio, TX.
3. Garrett W. Meigs et al., "Do Insect Outbreaks Reduce the Severity of Subsequent Forest Fires?," *Environmental Research Letters* 11, no. 4 (2016): 045008, doi:10.1088/1748-9326/11/4/045008.

THEN AND NOW, NOW AND TO COME

1. I want to thank Mike Lewelling for his written observations and for setting up a meeting at the park, even though he could not personally attend, and Nate Williamson and Doug Watry for a lively and informative tutorial in fire at Rocky.
2. The story Doug Watry told me deserves a wider audience. Wilderness advocates were unyielding about not allowing mechanical equipment,

so a contractor recruited a group from Latin America who did the work with hand tools and stacked the debris as they would charcoal mounds.

FATAL FIRES, HIDDEN HISTORIES

1. Wallace Stegner, *Where the Bluebird Sings to the Lemonade Springs* (New York: Modern Library Classics, 2002), 205.
2. Blackwater fire quote from David P. Godwin, "The Handling of the Blackwater Fire," *Fire Control Notes* 1, no. 7 (December 1938): 381.
3. See Glenn B. Stracher et al., "The South Cañon Number 1 Coal Mine Fire: Glenwood Springs, Colorado," *GSA Field Guides*, 2004, vol. 5, 143–50, doi: 10.1130/0-8137-005-1.143; and Sallee Ann Ruibal, "Coal Seam Fire Memories Still Burning a Decade Later," *Post Independent*, June 8, 2012, at http://www.postindependent.com/news/coal-seam-fire-memories-still-burning-a-decade-later/, and for a later update, John Stroud, "State Works to Control Burning Coal Seam in South Canyon," *Post Independent*, June 18, 2017, http://www.postindependent.com/news/local/state-works-to-control-burning-coal-seam-in-south-canyon/.
4. The technical report (and it is technical) is available at Howard A. Tewes, *Survey of Gas Quality Results from Three Gas-Well-Stimulation Experiments by Nuclear Explosions*, Lawrence Livermore Laboratory UCRL-52656, January 23, 1979. For a gentler summary of both fires, see Ken Lammey, "The 1976 Battlement Creek Fire—A Historical Perspective," Battlement Mesa Service Association, July 2012.
5. The fire has been the subject of a Wildland Fire Staff Ride, so a rich cache of documentation is available for it. See http://www.fireleadership.gov/toolbox/staffride/library_staff_ride10.html.

EPILOGUE: THE INTERIOR WEST BETWEEN TWO FIRES

1. Interestingly, a century later the writer John McPhee repeated that traverse, then extended its reach from New Jersey to California. His emphasis was, however, again on rocks—the new geology of plate tectonics. It would be interesting to see the traverse redone from an ecological perspective.

INDEX

Abbey, Ed, 88
Alaska, x, 17, 38, 86
American Indians, and fire, 7–8, 10, 103, 115, 120. *See also* Paiute forestry
Anthropocene, 39, 47, 62, 64–65, 68, 100, 145
Arapaho National Forest, 144–45, 151
Arid Lands. See *Report on the Lands of the Arid Region of the United States*
Arizona Strip, 69–78, 103
Arizona, 63, 69–70, 74, 103, 108, 156, 159. *See also* Arizona Strip
aspen, 112–15

Babbitt, Bruce, 74
Basin Range, 4, 13, 19, 33, 90, 100, 108–9, 165
Battlement Creek fire, 156, 161–63
Bel Air-Brentwood fire (1961), 53–54
Big Blowup, 3, 10, 45, 47, 91, 156–58
Black Forest fire, 137, 142
Black Tiger Gulch fire (1989), 136
Blackham, Leonard, 96
Boise, Idaho, 15, 19, 37, 43, 45–46
Bond, Horatio, 51
Boulder, Colorado, 5, 136, 139
Brand, Stewart, 58
bristlecone pine, 57, 61–63, 65–68, 109, 113
brome, 29, 32–33, 54–55. *See also* cheatgrass
Brown, A. A., 52
Bureau of American Ethnology, 9
Bureau of Land Management (BLM), 3, 16–18, 20–22, 26, 31, 37–38, 41–43, 46–48, 70, 72–74, 76, 109, 167
Burning Man, 13, 16, 27

CalFire, 25, 140
California, ix, 4–5, 11, 13–15, 19–20, 23–26, 29, 52, 86, 91, 94, 103, 133, 135–37, 140, 156, 164, 167. *See also* CalFire
Carson City, Nevada, 5, 13, 15–16, 19, 22, 26
Catastrophic Wildfire Reduction Strategy, 95–97
Catfire. *See* Catastrophic Wildfire Reduction Strategy
Center of Excellence for Advanced Technology Aerial Firefighting, 140, 142

cheatgrass, 5, 15–16, 18–19, 22, 29–49, 54–55, 63, 74–75, 79 fig. 1, 79 fig. 2, 91, 95, 100, 107, 109, 112, 115, 122, 126, 147, 162, 165, 167; large fires in, 37–38, 45–47; relationship to fire, 45–46
chukar partridge, 30, 36
Civilian Conservation Corps, 126, 158–59
Clark, Walter Van Tilburg, 17
Clementian theory of ecology, 32, 34–35, 153
Clements, Charlie, 30
Clements, Frederic, 10, 153
coal mine fires, 161, 163
Cohen, Michael, 66
ColFire. *See* Colorado Division of Fire Prevention and Control
Colorado Department of Public Safety, 140
Colorado Division of Fire Prevention and Control, 140–43; comparison to Southern California, 141–42
Colorado Plateau, 4, 42, 58, 85, 100, 108–9, 116, 123
Colorado Springs, Colorado, 4, 137
Colorado State Forest Service, 137–39
Colorado, xi, 4–5, 7–8, 10–11, 85, 93–94, 100, 133–44, 146–47, 156, 161, 163, 166; big fires, 136–37; profile of, 133–35
Comstock Lode, 13, 15, 17, 22, 26, 164
conservation (state-sponsored), 5, 8, 33, 88–90, 95, 109, 134, 136, 165–67
Covington, Wally, 74
crested wheatgrass, 41, 44–45
Curry, Donald R., 65–66

Denver, Colorado, 4, 133–34, 137
Department of Defense, 52–53
Department of the Interior, x, 5, 9, 47, 72, 74. *See also* Bureau of Land Management; U.S. Geological Survey; National Park Service; U.S. Fish and Wildlife Service
Deseret, 8, 13, 69, 85–86, 89, 92, 117, 122
Dutton, Clarence, 6, 71, 77–78

Elko fires (1964), 3, 16, 37
Elko, Nevada, 18, 26
Emigration Canyon, 97–98
Endangered Species Act, 36, 47, 165. *See also* sage grouse
exotic species. *See* invasive species

fatality fires, 156–63
Fire Temple, 129
fire, as weapon, 50–55
Flint Hills, ix, 3
Florida, ix, 3, 27, 133
forest fire labs, 52
fossil fuel fires, 161–63
Freeland, Joe, 26
Front Range (Colorado), 11, 93, 134, 136–44, 148, 152, 154, 166–67

Geographical and Geological Survey of the Rocky Mountain Region (Powell Survey), 5, 7–8, 10, 89, 91, 93, 165. *See also* John Wesley Powell; *Report on the Lands of the Arid Region of the United States*
Geological Exploration of the 40th Parallel, 164, 166
Giant Joshua, 87, 112
Gilbert, G. K., xi, 6, 93, 120, 165
Grand Canyon National Monument, 73
Grand Canyon National Park, x, 69–70, 72, 76
Grand Staircase, 77, 109, 115–17
Grazing Service, 37, 72. *See also* Taylor Grazing Act
Great Basin Experiment Station, 33, 38, 40
Great Basin National Park, 59. *See also* Snake Range
Great Basin Rangelands Research program, 33

Great Basin Restoration Initiative, 46–47
Great Basin, 4, 11, 13–19, 22, 27–40, 44–50, 54–56, 58–61, 85–86, 93, 100–101, 103–7, 112, 117, 122–23, 165, 167
Great Plains, 32–35, 105, 112
Greeley, William, 51
greenstripping, 44, 46

Henry Mountains, 119–21
High Park fire (2012), 137, 139, 146
High Plateaus (Utah), 108–16, 121–24; histories of settlement, 115–16
Hiroshima, 50, 84 fig. 8
Holmes, William Henry, 77–78
Holocene, 27, 62–63, 100, 110, 113–14
Holsinger, S. J., 103
Horse Pasture Mesa, 115–16
Humboldt National Forest, 26, 58
Humboldt River, 13, 15, 103

insects, compared to fire, 145–46
invasive species, 29–39; origins of, 27–29. *See also* cheatgrass; pinyon-juniper

James, William, 49
Jewell, Sally, 47, 166

Kaibab National Forest, 70, 73
Kaibab Plateau, 69–70, 77
Kansas, 40, 42, 44, 133
King Survey, 164–65
King, Clarence, 164
Kyle, Jon, 74

Lake Tahoe, 15, 17, 19
Landfire, 5–6, 90, 145
Las Vegas, Nevada, 14–17, 19, 69, 164, 169
Legarza, Shawna, 18
Leopold, Aldo, 30, 38–39, 41, 49
Lewelling, Mike, 155
Libby, Willard, 53–54
lightning fire, 7, 10, 15, 21, 28, 55, 57, 60, 62, 71, 73, 76, 94–95, 102, 110, 117–18, 120, 126, 139, 162, 166
Little Valley prescribed fire (2016), 26
Long Now Foundation, 57, 62–64, 68
Lower North Fork fire (2012), 137, 139

Maclean, Norman, 158–59, 163
managed wildfire, 119, 124
Mann Gulch fire (1949), 158–59
Mark Twain, 14, 17, 21, 87
McArdle, Richard, 158
McGee, W J, 90
Mesa Verde National Park, 125–131; big fires in, 126–27
Milford Flat fire (2007), 92, 95
Mill Flat fire (2009), 118
Miller, Richard, 101
Mormon Country, 123, 169
Mormons, 8, 13, 42, 69, 72, 85–89, 91, 93–94, 97, 109–10, 112, 117, 122–23, 133, 166; literature of, 87–88
Mount Trumbull Ecosystem Restoration Project, 74–78
Mount Trumbull Wilderness, 76, 81 fig. 4
Mount Washington, 58, 61–63, 67
Mount Wheeler, 58, 60–62, 65
mountain pine beetle, 144–49
Murphy Complex (2007), 47–48

narrative, characteristics of, 66–67; competing versions for High Plateaus, 115–16, 121–24
National Cohesive Strategy, 11, 26, 96, 140, 143, 154
National Fire Plan, 24, 26, 46, 152, 154
National Fire Protection Association, 51, 136
National Interagency Fire Center, 38, 43, 45

National Park Service, x, 59–60, 130. *See also* Great Basin National Park; Rocky Mountain National Park; Zion National Park
Nevada Cohesive Strategy, 26
Nevada Division of Forestry, 25–26
Nevada Test Site, 16, 54–56, 70
Nevada, xi, 4–5, 8, 10–11, 13–19, 21–22, 24–27, 31, 44, 50, 57, 65, 70, 72, 85, 86, 94, 100, 103, 117, 122, 133–34, 166–67, 169; profile of, 13–19. *See also* Great Basin; Sierra Front
Northern Rockies, ix, 3, 10, 91, 112
nuclear weapons, 52–55

Office of Civil Defense, 52, 54
Olson, Jim, 150–51
Ouzel fire, 136, 150–52

Paiute forestry, 10
Pando (aspen), 113–14
Parashant-Grand Canyon National Monument, 73, 76
Parley's Canyon, 98
Pellant, Mike, 40–49
Phillips Ranch fire (2000), 57, 60, 62, 67–68
Pickford, G. D., 33, 38, 40
Pinchot, Gifford, 9, 90
Pine Valley Mountains, 116–19
pinyon-juniper, 31, 33, 36–37, 41, 54, 60, 71, 74, 91, 100–107, 111, 120, 126–28, 131, 165, 167; controversy over, 106–7
Plateau Province. *See* High Plateaus
Pleistocene, 27, 39, 59, 61, 63–64, 100–101, 110, 114
Powell, John Wesley, 5–10, 89–91, 93, 95, 103, 165, 169. *See also* Geographical and Geological Survey of the Rocky Mountain Region; *Report on the Lands of the Arid Region of the United States*
Project Flambeau, 52
Project Mandrel Rulison, 161–62
Prometheus, bristlecone pine, 61, 65–68; myth of, 61, 65–66
Pyrocene, 38–39

railroad, Interior West, 28–29, 164–67
range science, 31–37
Red Butte Canyon, 97–98
Reno, Nevada, 5, 13–16, 19–21, 33, 40
Report on the Lands of the Arid Region of the United States, 5–6, 8–10, 89–90, 93, 167, 169
Roark, Kevin, 50
Rocky Mountain National Park, 136, 150–55
Rocky Mountains, 3, 8, 10, 15, 41, 64, 91, 94, 100, 108, 112, 136, 154. *See also* Northern Rockies
Roosevelt National Forest, 139, 144–46, 150–51
Roper, Bob, 26
Roughing It, 14, 17, 87

Saddle fire (2016), 118–19
sage grouse, 11, 16–18, 22–24, 26, 31, 38, 47, 96, 105, 148, 167
sagebrush steppe, 3, 14, 41, 102, 105, 165
sagebrush, 3–5, 14, 18–20, 29, 31, 33, 36, 41, 49, 54–55, 63, 79 fig. 1, 100, 111–12, 114, 165. *See also* sagebrush steppe
Salt Lake City, Utah, 15, 19, 88, 91, 94, 97–98, 117
Sand County Almanac, 38, 41
Sargent, Charles S., 5, 15, 20, 90, 101
Secretarial Order 3336, 23, 47. *See also* cheatgrass; sage grouse
Shantz, H. T., 33
Sierra Front Wildfire Cooperators, 24
Sierra Front, 19–20, 22–24, 26, 82 fig. 6
Sierra Nevada, x, 3–4, 8, 13–14, 19–22, 53, 94, 117, 164, 167. *See also* Sierra Front

Snake Range, 57–58, 60–61, 63–64, 67–68
Snyder, Gary, 57
South Canyon fire, 156–60
Southern California, 3, 14, 20, 23–24, 112, 117, 127, 131, 141
St George, Utah, 70, 72, 111, 117–18
Stegner, Wallace, 13, 90, 122–24, 157, 169

Taylor Grazing Act (1934), 16, 20, 37, 72–73
Texas, 27, 32, 72, 86, 88, 91, 133
Toroweap Valley, 71–73, 77–78
Trumbull Range, 71–78, 81 fig. 4, 103, 111

U.S. Fish and Wildlife Service, 47
U.S. Forest Service (USFS), x, 3, 14, 16, 18, 20, 23, 25, 33, 37, 40, 47, 51–52, 54, 59–60, 65–67, 72, 82 fig. 6, 98, 109, 111, 133–34, 142, 157, 165, 169
U.S. Geological Survey (USGS), 5, 9, 47, 71, 89–90, 93, 103, 164–65
Uinkaret Plateau, 69, 71, 76
Utah, xi, 5, 6–8, 10–11, 13, 41–42, 44, 69–70, 72, 79 fig. 1, 81 fig. 5, 83 fig. 7a, 83 fig. 7b, 85–100, 108–12, 122–23, 139, 165–66

Wasatch Front, 4, 40, 70, 85, 87, 93, 98, 109, 166. *See also* Wasatch WUI
Wasatch WUI, 93–99
Washoe Zephyr, 21–23, 26, 94
Whipple, Maurine, 87
Wigand, Peter, 101
wildland-urban interface (WUI), 19, 22–24, 26, 38, 46, 53, 93, 95–96, 125, 130, 141, 152. *See also* Front Range; Wasatch WUI; Washoe WUI
Wilson, Carl, 162
Winnemucca, Nevada, 17, 26, 31, 103

Young Men and Fire, 158–59
Young, Brigham, 87, 89, 97
Young, James, 30

Zion National Park, 77, 79 fig. 2, 109, 115–16

ABOUT THE AUTHOR

Stephen J. Pyne attended Stanford University and the University of Texas at Austin. He is currently Regents Professor in the School of Life Sciences at Arizona State University. Among his recent books (of over 30 published) are *Voyager: Seeking Newer Worlds in the Third Great Age of Discovery*; *The Last Lost World: Ice Ages, Human Origins, and the Invention of the Pleistocene* (co-authored with Lydia V. Pyne); *Fire: Nature and Culture*; and his multivolume survey of the contemporary American fire scene for the University of Arizona Press: *Between Two Fires* and the To the Last Smoke series. He is a former North Rim Longshot.